# POSTULATION, ELABORATION AND APPLICATION OF

# EXTRAABSTRACT LANGUAGES

## 2nd Edition

### PART 1:
POSTULATES OF EXTRAABSTRACT LANGUAGES
SYNTAX OF MATRICAL SITUARICS

### PART 2:
TRANSFORMATION OF MATRICAL SITUARICS INTO
RECOMPLEX TENSOR SEQUENCE
DEVELOPMENT OF A PARADIGMATIC ALGORITHM

### PART 3:
PARTS SOURCE CODE

## WALDEMAR SCHWARZKOPF

'[...] Until the rise of a visionary new culture that once again embraces the cosmic perspective; a perspective in which we are one, fitting neither above nor below, but *within.*'

- Neil deGrasse Tyson

'Your assumptions are your windows on the world. Scrub them off every once in a while, *or the light won't come in.*'

-Isaac Asimov

# Content

PART 1: POSTULATES OF EXTRAABSTRACT LANGUAGES;
SYNTAX OF MATRICAL SITUARICS

**PART 3: PARTS SOURCE CODE**

# Preface [1 - 5]

If one was asked to describe human nature, one of the most important properties of humans to be mentioned would be their curiosity. In fact, we are not the only species to be regarded as curious. Nevertheless, we developed a far superior version of it. Our extensive curiosity not only led us to survive, but made it possible to spread out into the world and, at least partially, beyond.

As *curious* beings, we steadily ask: 'Why?' As *intelligent* beings, we try to find the answer by searching for patterns and explain it by hypotheses. First, there were myths, from which religions emerged. Later on, our curiosity forced us to ask questions with ever-increasing precision. Accordingly, explanations were required to increase their precision as well, therefore logics, mathematics and science appeared. Over the centuries, our species developed a steadily increasing momentum of knowledge gaining.

Nowadays, digitalization and globalization have social, economic, and political impacts on our civilization. The omnipresent access to information and worldwide communication does not only allow connecting and educating oneself. Unfortunately, the sheer mass of information implies the difficulty of differentiation between verified facts and 'fake news', which spread and mutate globally, especially due to social media. Furthermore, digitalization allows new forms of economic and political exertion of influence. It follows that these developments have advantages and disadvantages as any other in human history. But after witnessing the effects of colonialization and industrialization, after becoming responsible for two devastating world wars and increasing the self-made risk of extinction by developing weapons of mass destruction, we should be more aware of our responsibility when facing a multivalent development of our global civilization. At least more than we are currently.

Indeed, possibilities of globalization and digitalization shall not remain unused, since they are actual chances to evolve into a civilization of type one on the Kardashev scale, which will hopefully decrease the chance of extinction. But we must admit that we still are gregarious animals. This property of our species leads to conflicts, which may end up lethal for the whole species, given our gained knowledge and creative minds. Thus, it is required to use every chance with respect and responsibility. In order to have an oversight about all the manifold developments, we need to classify them. Hence, we require a general system, in which we have a clear sight which is crucial for a respectful and responsible course of action. As a communicative species, our systematization begins with language. Consequently, we need a language, in which we can sum up all knowledge, every observation and any kind of processes, especially decision-making. For instance, the development of artificial intelligence and the dawn of expansion into space require clear, fast and sustained decisions. Moreover, climate change and, as we shortly witnessed, pandemics and redundant racial conflicts remind us not only to act together as one species, but also as a *truly intelligent* one.

The following work is the result of a process which began with a very simple question: 'How to describe everything in the same manner?' After numerous approaches, which sooner or later turned out to be defective, an axiomatic system for any all-encompassing language could be derived. Any mentioned theoretical development has been chosen to explain the capability of the axiomatic system and suggest new approaches to current research fields.

Other, insufficiently developed approaches and theories have been either dismissed or are mentioned shortly as a thought-provoking impulse.

## Acknowledgements

First, I would like to thank Hermann Bauer and Elmar Schönecker for teaching me essential mathematical methods and philosophical perspectives, from which the initial motivation emerged. Furthermore, I would like to mention that the University of Saarland provided both educational and logistic prospects for elaborating and writing, especially Dr. Cord Friebe, Dr. Michael Springborg and Dr. Valeri Grigoryan. Moreover, I am thankful for the great help I received from Irene Villalba Güemes, Michael Dorongovski, Leonardo Fabela, Sven Ehses and, of course, my family.

# PART 1:
## POSTULATES OF EXTRAABSTRACT LANGUAGES
## SYNTAX OF MATRICAL SITUARICS

# 1. Postulates of the Extraabstract Languages [6 - 13]

## 1.1 Clarification of the Term 'Everything'

The intention of this work is to create a language which is capable to describe everything. The term 'everything' can be interpreted in various ways, but for the purposes of this work, the most naive and simple interpretation has been chosen. Therefore, everything means anything describable, imaginable, and predictable and, consequently, anything beyond these subjective restrictions. The consequences of this interpretation will be illustrated by definition and elaboration of such a language. First, it has to be made clear *why there is a requirement* for it at all.

### 1.1.1 Current Situation in Science

After renaissance, science has emerged to an essential branch of human activities, especially in terms of development of technologies and improvement of living conditions, and, following our nature, also in warfare. Industrialization, two World Wars, and the Cold War followed scientific development and vice versa. The interdependence lead to an ever-increasing sophistication in physics, chemistry, medicine and engineering, as well as in other many branches which satisfy the attribute of paradigmatic development and thus can be regarded as scientific fields. Especially the desire to become militarily and economically invincible made us gather knowledge at a much higher rate than ever before. Our 'good will' to deal with the consequences of past, imprudent actions and inventions led to new technologies and applications as well, though.

However, with more knowledge, one needs more scientific branches, as it allows proper organization. A significant example is provided by Newtonian celestial mechanics and Einstein's theory of relativity: Whereas Einstein proved Newton to be wrong *in general*, we still use the classical equations to send probes into space to reach the edge of our solar system. But even when it comes to trivial applications like GPS, we need to include Einstein's equations to calculate positions accurately. That is, as long as the circumstances allow, we can apply Newton's equations to predict events and calculate properties with a sufficient accuracy, but as soon as relativistic effects become significant, we must use

1

Einstein's equations. Hence, Newtonian laws can be interpreted as specialized approximations of Einstein's theory.

A far more crucial example includes Einstein's theory of relativity on the one hand, and quantum mechanics on the other hand. Whereas the relativistic perspective includes events on large scales, quantum mechanics discusses events on the microscopic scale. Moreover, when drastic circumstances force us to apply both branches, like the strong gravitational forces caused by black holes, both theories give different predictions. In fact, both general relativity and quantum mechanics occur to be *incommensurable*. Their models describe the world in two different ways, which leads to different predictions and thus a break of scientific consistency.

Another case of incommensurability occurs when we compare physics to chemistry or psychology. As the older scientific field, physics includes far more advanced mathematical instruments than other scientific fields, both qualitatively and quantitatively. Let us say, physics is more *mathematized*. Chemistry, however, evolved centuries later and is less mathematized, whereas every branch (organic chemistry, analytical chemistry, etc.) appears to have a different grade of mathematization. Psychology, as the youngest of the three named scientific fields, appears to have the lowest grade of mathematization. It seems to be the inevitable fate of sciences to be mathematized as they evolve to increase their precision. The incommensurability emerges between low-mathematized and high-mathematized scientific fields, since their language, or *syntax* differs alongside the difference of their mathematization grade.

Hence, incommensurability between scientific fields makes it difficult or even impossible to fuse them together. As our knowledge increases as well as the amount of scientific branches, we face a divergent trend in science. In fact, there are numerous advantages in fusing scientific branches, as we have seen in bionics, for instance. To benefit from these advantages, we must overcome incommensurability. This means, we require a language, which unifies any syntax used by the different branches, not only in terms of mathematization, that is precision, but also in terms of different perspectives.

### 1.1.2 Requirements for an Abstract Language

To overcome the problems depicted in 1.1.1), we require a language which can describe *anything* in the same manner. This means, that not only scientific or mathematical problems shall be able to be formulated, but also subjective views, experiences, feelings and nonsense as well. As has been said before, 'everything' shall be interpreted in the simplest way: The abstract language has to be able to describe everything, independent of physical or mental restrictions.

The consequence is a general consistency of syntax – it has to remain the same independent of what is to be described. By syntax consistency, incommensurability can be inhibited. Speaking of mental restrictions, an abstract language has not only to describe anything we can imagine, but also things beyond this border. It must be able to describe the inconceivable, or even the indescribable, since we shall not limit ourselves to things which are mentally accessible. The motivation for this requirement can be illustrated by the quote of John B.S. Haldane: '[...] the universe is not only queerer than we suppose, but queerer than we *can* suppose'. [14]

Indeed, this requirement may lead to paradox-like statements, but as we intended to describe everything in its most simple meaning, a paradox shall be able to be described as well.

Another requirement is the ability to vary the precision. This is essential when we need to formulate approximations which enable scientists to elaborate useful scientific instruments without complex formulation. Furthermore, it enables computer programs to work more efficiently. Calculations with great precision require more time, energy and thus financial resources. Given the ability of variable approximation grade, a language establishes the ability to increase its own efficiency.

Another requirement can be derived from the formulated ones: The abstract language has to be able to describe itself and its own descriptions. This is significant in context of derivations of new laws which have an abstract manner, and, more importantly, reformulating the language. Moreover, this can lead to more precise descriptions of consciousness and thus a new kind of artificial intelligence.

Summing up, the requirements can be formulated in a property which defines the purpose of this work: Oversight with insight.

In order to formulate the axiomatic system for languages which satisfy the depicted requirements, we need to formulate crucial differentiations and definitions first.

## 1.2 Extraabstract Languages

### 1.2.1 Extraabstracticity

First, we need to differentiate levels of abstracticity. To do so, we classify three different spheres, which include things and events of the same level of abstracticity. Afterwards, we evolve a new dimension, as we need a theoretical basis which is independent of abstracticity levels. This dimension must include the type of language of describing all the previous levels with a consistent syntax.

### 1.2.2 Sphere Model

#### 1.2.2.1 Subabstract Sphere

The subabstract sphere includes all things and events which are not abstract in the classic way of understanding. This means, naively spoken, things we can touch and events which extent is limited by time. In fact, everything with a space-limited extent is limited by time as well. If we ignore the differentiation of things and events in favor of general formulation, we need to clarify the difference between solely time-limited and space-limited things. For instance, the communication of neurons in the brain are space-limited (and thus time limited as well), but the resulting thoughts are time-limited, as thoughts do not have an extent in space by definition. Hence, the subabstract sphere contains two levels, the time level with higher rank and the space-time level with lower rank.

## 1.2.2.2 Abstract Sphere

The abstract sphere includes, first of all, scientific theories which make predictions and allow us to calculate properties of elements of the subabstract sphere (whether the predictions and calculations are right, does not matter at this point). These theories have an abstract nature since they can be regarded as timeless. Their limitation does not depend on space or time, but on their application and mathematical rules. Thus, the lowest level of the abstract sphere represents scientific theories which describe limited amounts of the subabstract sphere and the specifications of the higher level, the mathematical one. Since mathematics is limited by the laws of logic, this is the third and highest level of the abstract sphere. It is worth to mention that every limitation provides a range of consistent syntax. For instance, mathematical limitation provides a consistent syntax for all mathematical formulations. However, this syntax cannot be applied for non-mathematical formulations.

## 1.2.2.3 Metaabstract Sphere

As a space-time being, it is hard to clarify whether there is a metaabstract sphere or not. In fact, the formulation of the postulates does not depend on the existence of this sphere, but since we intend to describe also illogical events, we may assume that it does exist. Whatever the nature of the elements of this sphere may be, the lowest level of the metaabstract sphere must define the limitations of logics, as it is the highest level of the abstract sphere. Since it is not required for this work to formulate a concrete structure of the metaabstract sphere, we may simply assume that there could be such one, or even more of them. Whether they exist and which structure they in fact have, is the matter of further research.

## 1.2.2.4 Extraabstract Sphere

Independent of the amount of metaabstract spheres (ranging from zero to an arbitrarily high number) we can define the first abstracticity grade as the value of first abstracticity dimension, which can be manifested by an axis. The second abstracticity grade may be the second dimension, manifested by an orthogonal axis with respect the first one. In this dimension, the extraabstract sphere is defined as an abstracticity grade above the first axis'

4

abstracticity grade. Thus, this sphere provides the limitation to all the spheres on the first axis. Within this limitation, the syntax is consistent, which means that every first abstracticity level can be described using the same syntax.

This provides the theoretical basis for a language satisfying the requirements stated in 1.1.2. Thus, the postulates for the *extraabstract languages* can be formulated.

## 1.3 The Postulates of Extraabstract Languages

*0)*     Descriptiveness implies predictability. The accuracy of predictability depends on the grade of descriptiveness.

*1)*     Everything can be described. Nothing is excluded – thus, there is nothing indescribable. According to *0)*, everything can be predicted with a given accuracy, if the description is sufficiently accurate.

*2)*     For the purpose of a description, everything can be seen as an *extraabstract situation*, including the description itself. There are infinitively many extraabstract situations, and, infinitely many ways of descriptions. According to *1)*, all ways of descriptions are describable.

*3)*     A *language* is a way of information reorganization, which provides the ability to describe. According to *2)*, a language is an extraabstract situation and there are infinitively many languages. It is possible to formulate a language which is capable of describing everything according to *1)*. Then it is capable of describing itself as well.

*4)*     A language described by *3)* is concretely formulated. This means, it resembles a special syntax.

*5)*     Each language characterized by *3)* and *4)* can be transferred into another of that type. According to *1)*, *2)* and *3)*, such a language can be described by another one. Hence, the transformation can be realized by the starting language itself.

It is important to note that according to the principle of relativity, the first axis also provides the limitation to all the spheres on the second axis as well. This consequence is plausible: As the following chapters will show, we must, as space-time beings with the ability to formulate mathematical descriptions, apply mathematical and logical language in order to describe the extraabstract sphere and thus any language of extraabstract nature. The abstract sphere provides *our* limitation of description, and we are obliged to operate within this limitation.

However, the postulates provide the basis for language construction according to 1.1.2. In this part of the work, the syntax of such a language will be elaborated. In PART 2, the language will be transformed into another one as an example for the fifth postulate. Furthermore, it will be applied on a current problem in science theory. PART 3 contains the translated solution of the problem as a paradigmatic algorithm. This represents a blueprint for an autonomous development of scientific theories and the prospect of total scientific unification.

## 2. Basics of Matrical Situarics [12, 13, 15]

The most general differentiation we need to undertake outlines the most fundamental instruments of the language. As generality also includes simplicity, the first differentiation must be as simple as possible.

### 2.1 Differentiation between Situations and Non-situations

An event includes things which interact with each other. Whether they interact and the type of interaction depends on their properties. For instance, the gravitational interaction solely occurs between objects with mass (or energy, in general), whereas magnetic interaction solely occurs between objects with a charge. For instance, ideas cannot interact in both ways, since they have neither mass nor energy nor a specific charge.

In the next step, we split the concept of interactions into affects. Each affect belongs to a property which again belongs to a thing. If two things interact with each other, this means their affects aim for the correspondent other thing and belong to the same type, such as gravitational or magnetic effects.

'Things' in general can be of different types. As the previous chapter has shown, there are different spheres and levels which define the types of things. In fact, the definition of types does simply distinguish between different properties, so things without properties can be regarded as completely unspecified. For instance, the constellation Ursa Major, represents a bear in Greek culture. In Roman culture, the stars represent seven bullocks wandering around the north.

In our terms, we can define a 'thing' as a *finality*. Finalities are carriers of properties which define according affects. This kind of syntax is well known: We use it in our languages as well. Finalities are nouns, properties are adjectives or attributes in general, and affects can be regarded as verbs. The sentence is the equivalent to the described event itself, which includes the depicted building blocks.

According to the postulates, every building block of this language can be regarded as a situation. The main difference between finalities and the other building blocks is that a finality is unspecified. This means, the finality is the most general thing, thus it represents the end of any generalization or approximation range. For instance, we can formulate the interdependence between weather, geology, fauna and flora on earth as a situation according to the postulates. However, usually, we simply refer to all of it as 'nature'. In chapter three, the exact formulation of both generalization and approximation will be depicted. Since the *final* generalization or approximation is represented by the finality, its notification is appropriate. Furthermore, because the other building blocks can be specified, thus described in a more accurate way, we may call all of them *prime situations*, because *in the beginning*, we must assume them to be describable situations. The event, which includes the other building blocks, may be called *matrical situation*. Instead of 'property' we will use the term *attribute*.

Hence, we have four different building blocks, a finality and three prime situations. Finalities are written as *Fin*, prime situations as *Sit*. To distinguish prime situations, they are classified with Greek letters. Summing up we write:

*Finx*    for the finality called *x*

*Sit$\eta_x$*    for the attribute of finality called *x*

*Sit$\varepsilon_x$*    for the affect of finality called *x*

*Sit$\sigma$*    for the matrical situation in which the other building blocks are embedded.

These building blocks will be used to construct the syntax of the language, which is called *matrical situarics*, hereinafter MS. Following, the syntax of MS will be formulated as well as basic elements which will be evolved within this part.

## 2.2 Syntax of MS

We can formulate the matrical situation *Sit$\sigma$* as a matrix, in which each row is reserved for a building block each. The first raw from above contains affects, the second attributes and the third finalities. Each column describes a finality with its according attributes and affects. The matrix is written with curly brackets after the according symbol of the matrical situation. We may call the matrix itself as *situative matrix*, whereas all elements within it are called *situative elements* or simply *situatives*. It is allowed, however, to leave out spaces for affects, attributes or finalities. The reason is that we do not always know which affect is defined by a known attribute, or which attribute belongs to a given finality, for instance. In fact, we can also have solely the information of the affect or the attribute, without knowing to which finality they belong. The simplest structure as well as a matrix including every of those cases is shown in (2.1) and (2.2), respectively.

$$(2.1) \qquad Sit\sigma := \left\{ \begin{array}{c} Sit\varepsilon_1 \\ Sit\eta_1 \\ Finx_1 \end{array} \right.$$

$$(2.2) \qquad Sit\sigma := \left\{ \begin{array}{ccc} Sit\varepsilon_1 & Sit\varepsilon_3 & Sit\varepsilon_5 \\ Sit\eta_1 & Sit\eta_2 & Sit\eta_4 \\ Finx_1 & Finx_2 & Finx_3 \end{array} \right.$$

If two finalities share the same attributes, and thus same affects, but still are independent finalities on their own, we can sum them um in squared brackets. These are called *double occupations* or *multiple occupations*. It is worth to mention, that two different finalities must have different attributes in order to be distinguished. Otherwise, they would turn out to be the same thing. However, since we can describe situations which only refer to special attributes (and affects), we can leave out the mutual attributes if we are not interested in them. If we observe two objects with different influences on a given observed system, we do not care about their mutual attributes, as we focus on the differences: For instance, if two planets have different gravitational effects in their planetary system, it is not important that they have the same shape or color, or that they both are, *per definitionem*, planets. A simple example of double occupation is shown in (2.3).

$$(2.3) \qquad Sit\sigma := \left\{ \begin{array}{c} \left[ \begin{array}{c} Sit\eta_1 \\ Sit\eta_2 \end{array} \right] \\ \left[ \begin{array}{c} Finx_1 \\ Finx_2 \end{array} \right] \end{array} \right.$$

Whenever we change the position of any element within a situative matrix, we change the matrical situation by itself. We may call this the principle of *integrity*. A set of given situations, affects, attributes and finalities with given integrity is called a *situaric system* or simply *system*. The integrity of a situation is always the same, and a change in its integrity means a change of the situation and hence, of the whole system. We can differentiate between *intrinsic integrity* and *extrinsic integrity*. Intrinsic integrity is defined by the constellation of situative elements. Extrinsic integrity is defined by all elements outside the situative matrix, which will be discussed in the next chapter.

## 2.3 Introduction of Relations

### 2.3.1 Syntax and Relation Vector

In order to formulate statements about the situative matrix and thus increase the amount of information the language can contain, we need to be able to connect situative elements with each other. The simplest type of connection is a *relation*. We can write the relation as a tuple, in which the first element is the subject and the second one the object. They are separated by a semicolon, as shown in (2.4).

$$(2.4) \qquad \langle subject\,;object \rangle$$

According to the postulates, the relation is a situation too. We can formulate it with a situative matrix as well. This means, by changing the role of the according elements, i.e. subject and object, we obtain a new situation, thus relation, since the integrity changes as well. Furthermore, a relation can consist an arbitrary amount of elements between the subject and the object. All of them are objects of type two. Type one objects are always in the end of a tuple.

If we formulate a situative matrix with given integrity and intend to describe a relation between to situative elements, it is helpful to have an orientation. For this purpose, each relation has its own relation vector. This vector is two-dimensional, and consists of the steps one has to make starting from the subject to reach the object (or the objects of type two, in the according order). The path is visualized if we call the horizontal axis the x-axis and the vertical axis the y-axis and put the matrix into this coordinate system, whereas the first finality space lies in the origin. Since it is helpful not to count the steps, but the over leapt elements (or the empty spaces), the vector shows the amount of leaps. If the relation step reaches solely to the next space, we simply write a '+' or '-'.

If a relation has multiple steps, the vector is replaced by an addition of vectors, which are summed up in a matrix. If one intends to find out the 'shortcut' between the subject and one of the objects, one has to count all vectors together, which lie between these two elements into the relation vector. An example is given in (2.5). In this case, *R1* and *R2* are relations between the same elements, *Finx₁* and *Finx₂*, but *R2* describes the relation in detail, as it shows the exact path of the relation. Furthermore, *R2* and *R3* may be called *anti polar* with respect to each other, because their subjects and object are switched, whereas their paths remain the same.

$$(2.5) \qquad Sit\sigma := \begin{Bmatrix} \begin{vmatrix} Sit\varepsilon_1 & Sit\varepsilon_2 \\ Sit\eta_1 & Sit\eta_2 \\ Finx_1 & Finx_2 \end{vmatrix} \end{Bmatrix}$$

with

$$R1 := \begin{pmatrix} + \\ 0 \end{pmatrix} < Finx_1 ; Finx_2 >$$

$$R2 := \begin{pmatrix} 0 & 0 & + & 0 & 0 \\ + & + & 0 & - & - \end{pmatrix} < Finx_1 ; Sit\eta_1 ; Sit\varepsilon_1 ; Sit\varepsilon_2 ; Sit\eta_2 ; Finx_2 >$$

$$R3 := \begin{pmatrix} - \\ 0 \end{pmatrix} < Finx_2 ; Finx_1 >$$

## 2.3.2 Viality and Vial Aspect

When we formulate relations, we can also 'draw' a path through the coordinate system in which we imagine the situative matrix to be embedded. This path may be called *viality* [via=(lat.) way] and is characterized by the relation vector. There are *direct* and *indirect* vialities. Direct vialities occur, if all objects of type two are elements of the columns of the subject and the object of type one. Indirect vialities occur, accordingly, if at least one of the objects of type two is not element of either the subject or object type one column. Hence, relations without any objects of type two have automatically direct viality.

In case of double or multiple occupations, we need to distinguish which elements within the squared brackets are actually connected by the relation. An example is given by (2.6):

$$(2.6) \qquad Sit\sigma := \begin{Bmatrix} \begin{vmatrix} \begin{bmatrix} Sit\eta_1 \\ Sit\eta_2 \end{bmatrix} & \begin{bmatrix} Sit\eta_3 \\ Sit\eta_4 \end{bmatrix} \\ \begin{bmatrix} Finx_1 \\ Finx_2 \end{bmatrix} & \begin{bmatrix} Finx_3 \\ Finx_4 \end{bmatrix} \end{vmatrix} \end{Bmatrix}$$

If we formulate a relation between these four finalities with a given viality, there are different possibilities between which elements the relation actually occurs:

-between $Finx_1$ and $Finx_3$,

-between $Finx_1$ and $Finx_4$,

-between $Finx_2$ and $Finx_3$,

-between $Finx_2$ and $Finx_4$.

10

Whereas the viality is the same, we have now different combinations of subjects and objects. To include this variation into our syntax, we introduce the *vial aspect*. Dependent on the amount of possible subject and object occupations we choose, we can have three types of vial aspects. The *monovial aspect* describes the situation, if we solely choose one of the four possibilities (*monos*=(grk.) alone). *Compluvial aspects* are those in which more possibilities are considered (*complere*=(lat.) to fill). The *omnivial* aspect describes the consideration of all possibilities (*omnis*=(lat.) everything). The *nonvial aspect* occurs, if none of the possibilities are considered. One can argue that a relation with nonvial aspect simply does not exist. But for later analysis, it is important to be able to postulate relations and verify their actual existence afterwards. In case of mono- or compluvial aspects, we need to declare which elements are considered to be a part of the relation. To keep the formulation simple, we write a symbol, a number or a letter, representing each possibility to the according element. In case of omnivial aspects, it is sufficient to write 'omni' to the formalism of the relation. As an example for (2.6), (2.7) shows a mono-, complu- and omnivial aspect of the same relation.

$$
(2.7) \qquad R_{1,4} := \begin{pmatrix} + \\ 0 \end{pmatrix} \left\langle \begin{bmatrix} Fin_1^a \\ Fin_2 \end{bmatrix} ; \begin{bmatrix} Fin_3 \\ Fin_4^a \end{bmatrix} \right\rangle
$$

$$
R_{1,2,4} := \begin{pmatrix} + \\ 0 \end{pmatrix} \left\langle \begin{bmatrix} Fin_1^a \\ Fin_2^b \end{bmatrix} ; \begin{bmatrix} Fin_3 \\ Fin_4^{a,b} \end{bmatrix} \right\rangle
$$

$$
R_{omni} := \begin{pmatrix} + \\ 0 \end{pmatrix} \left\langle \begin{bmatrix} Fin_1 \\ Fin_2 \end{bmatrix} ; \begin{bmatrix} Fin_3 \\ Fin_4 \end{bmatrix} \right\rangle
$$

$R_{1,4}$ describes a relation solely between $Finx_1$ and $Finx_4$, thus it can be considered as monovial. $R_{1,2,4}$ appears between $Finx_1$ and $Finx_4$, but also between $Finx_2$ and $Finx_4$. Accordingly, this is a compluvial relation. $R_{omni}$ appears between all members of the subject-object pair as an omnivial relation.

### 2.3.3 Types of Situative Relations

Since all the relations we have discussed occur within the situative matrix, we may call them *situative relations*. A relation which elements are solely situative elements, is a situative relation.

Another classification, however, resembles the amount of information of a relation. If we formulate a situation with relations between the elements, these relations are part of the basal information about the situation. If we change the relation (for instance, by research), the change in information changes the integrity of the relation. One can say, the relation mutates. We can formulate it by applying a function on the basal relation which provides the mutated relation we obtain after research. Let the function be called µ-function. All basal situative relations have the formalism *A*, all mutated ones the formalism *B*.

Furthermore, we need to distinguish relations with respect to their elements. The most basal relation is called the *α*-relation. This is the relation between a finality and its attribute, without any objects of type two. Any other relation without objects of type two is called a *β*-relation. Finally, all relations with at least one object of type two are called *γ*-relations. If one wants to inform shortly about how many objects of type two the relation consists of, one can write their amount above the *γ*.

In chapter four, further types of relations will be discussed. The situative relations are the simplest, and the most important ones, since most description occurs within the situative matrix. This is why it is crucial to introduce them right in the beginning. Before the next level of description will be discussed, the structure of the syntax has to be narrowed first.


## 3. Situaric Space Model [16, 17]

### 3.1 Definition of Spaces

As we have stated in the previous chapter, the MS consists of finalities, attributes, affects and matrical situations as well as relations. To evolve our precision of description, we need to introduce a fabric into the language syntax. This can be realized by defining different spaces, which enclose different levels of description. The simplest one is defined by the matrical situation itself.

### 3.1.1 Situative Space

All elements within the situative matrix describing the situative matrix are called situative elements or just situatives. Relations within the matrix are situative relations. Accordingly, the interior of the situative matrix may be called *situative space*. This means, every matrical situation defines its own situative space. Spaces in general are symbolized by Cyrillic majuscles. The symbol for the situative space is a Cyrillic voiced 's': З.

### 3.1.2. Situatoric Space

As matrical situations are different from the building blocks they consist of, they need to be embedded into another space which contains the situative spaces defined by the matrices. To recall a physical pendant of this principle, I refer to the mathematician Theodor Kaluza, who wrote in 1921 that the universe may have more than the familiar three dimensions of space, whereas the additional dimensions are coiled up in microscopic ranges. We may call the engulfing space *situatoric*. Accordingly, matrical situations may be called *situatoric elements* or just *situators*. The symbol is a Cyrillic unvoiced 's': C. The peculiarity of the situatoric space is, that it can 'bend' into itself. If we describe a situation as an event, in which other events take place, like the movement of different parts of a Swiss clock, the events within the described event are also matrical situations, thus the situatoric space contains itself, or *bends* into itself. This iteration can be described by different *meta-levels* of the situatoric space and can be declared by exponents of the situatoric symbol, i.e. $C^2$, $C^3$ and so on. How the process of bending can be formulated in detail will be discussed in 3.3.

### 3.1.3 Situant Space

As we intend to describe events more precisely, we must consider, that finalities consist of other finalities, like a school consistsing of different grades. In fact, this consists-of-relation can be iterated an arbitrary amount of times. Given the example, a grade may consist of several classes, and each class consists of a given number of students. This simple example shows that the situative space must have more than two dimensions, which occur, similarly to Kaluza's theory, at smaller scales. In terms of MS, smaller scales mean higher precision. To maintain the simplicity of two dimensions, which enables us to formulate down the situative space on two-dimensional media like paper or screens without further complications, we introduce the *situant space*. Its elements are called *situant elements* or just *situants*. The situant space describes an additional, smaller space within the situative space. Situant spaces are defined by situatives (finalities, attributes, affects), similarly to the situative space. Different to situative spaces, situant spaces are automatically connected to their counterparts within the same column: The intrinsic integrity of the situants of a finality are similar to the intrinsic integrity of its attribute's situants, and accordingly to the intrinsic integrity of its affect's situants. This principle may be called *situant symmetry*. It is quite important as it allows us to change the intrinsic integrity on solely one situant within a column, since the other situants will change automatically. Furthermore, it enables regularity of the later defined situant relations.

For declaration, we use parentheses, which are written after the according, defining element, as **Fig.3.1** shows. The defining elements are called *meta-elements* or *metaelements*, whereas situants of a meta-element are called *sub-elements* or *subelements*.

Furthermore, the situant space bends into itself, like the situatoric space. This means that all parts of a situant remain to be situants. One can classify the different 'depths' of situant spaces as grades, similarly to the meta-levels of situatoric space. The symbol for the situant space is a Cyrillic short 'tsh': Ч.

### 3.1.4 Situaric Space

To embed all spaces into an all-engulfing fabric, we define the *situaric space* as the general space which contains the three minor spaces. This concept appears to be inescapable, since all elements are enclosed from each other by being put into different spaces which do not intersect. Since we are also interested to identify relations between elements of different spaces, we need a general fabric in which all elements occur. Thus, all situants, situatives and situators are automatically *situaric elements* or simply *situars*. The according symbol is a Cyrillic 'z': Ц.

To sum up the concept of the situaric space model, **Fig.3.1** shows an example of formulation with the differentiation of the spaces.

Elements in general have the formalism $E$, whereas their according space formalism is written as an index. If no special space is known or if we consider elements of different minor spaces, the situaric symbol can be written as an index.

**Fig. 3.1:** The different spaces. All elements are automatically elements of the situaric space. The according element symbols are following: For $Sito_R$: $E_Ц$, for $Finx_1$, $Sit\eta_1$, $Sit\varepsilon_1$, $Sit\eta_1$: $E_3$, for $Sit\eta_{21}$, $Sit\eta_{22}$: $E_ч$, for $Sit\eta_{211}$, $Sit\eta_{212}$: $E_{ч2}$.

$$Sito_R := \left\{ \begin{array}{c} Sit\varepsilon_1 \\ Sit\eta_1 \ Sit\eta_2 : \left( Sit\eta_{21} : \left| \begin{array}{c} \left( \begin{array}{c} Sit\eta_{211} \\ Sit\eta_{212} \end{array} \right) \\ Sit\eta_{22} \end{array} \right. \right) \\ Finx_1 \end{array} \right\}$$

## 3.2 Space Parts

By describing an event, we may be interested solely in few elements within given spaces. Given the example form 3.1.3., we may intend to formulate the relation between two siblings, who are students in the same school, but in different classes of different grades. Let them be named Max and Mary. In order to accentuate our interest towards several elements within a space, we define *space parts* as quantities of elements within spaces. The symbol for space parts is $A$. We can describe them as integrals with lines on each side. On the left side from the integral, the type of elements is shown. If all three types of minor space elements occur, we can use the formalism for situars. On the right side of the integral, we can write the constructed situations the elements are elements of the depicted space part.

These will be discussed in chapter four. According to their definition, space parts can be formulated as quantities of elements.

$$(3.1) \qquad A = E_{\text{Ц}} \left\| \int \right\| Konx_i = \{ \dots \}$$

Space parts allow choosing elements from different minor spaces. Let us assume that the school we were talking about in the example is a primary school with four grades. To keep the example simple, each grade may have two classes and each class 20 children. Max may be the 17th student in alphabetical order in class 1.1. (first grade, first class). Mary may be the 15th student in alphabetical order in class 3.2. (third grade, second class). The whole constellation can be formulated by (3.2) and the space part Max and Mary are part of can be written as (3.3).

$$(3.2) \qquad Sit\sigma_{School} := \left\{ \begin{array}{c} Sim_1 : \left[ \begin{array}{c} Sim_{1.1} : \left( \begin{array}{c} Sim_{1.1\,1} \\ \dots \\ Sim_{1.1\,20} \end{array} \right) \\ Sim_{1.2} : \left( \begin{array}{c} Sim_{1.2\,1} \\ \dots \\ Sim_{1.2\,20} \end{array} \right) \end{array} \right] \quad \dots \quad Sim_4 : \left[ \begin{array}{c} Sim_{4.1} : \left( \begin{array}{c} Sim_{4.1\,1} \\ \dots \\ Sim_{4.1\,20} \end{array} \right) \\ Sim_{4.2} : \left( \begin{array}{c} Sim_{4.2\,1} \\ \dots \\ Sim_{4.2\,20} \end{array} \right) \end{array} \right] \\[2em] Finx_1 : \left[ \begin{array}{c} Finx_{1.1} : \left( \begin{array}{c} Finx_{1.1\,1} \\ \dots \\ Finx_{1.1\,20} \end{array} \right) \\ Finx_{1.2} : \left( \begin{array}{c} Finx_{1.2\,1} \\ \dots \\ Finx_{1.2\,20} \end{array} \right) \end{array} \right] \quad \dots \quad Finx_4 : \left[ \begin{array}{c} Finx_{4.1} : \left( \begin{array}{c} Finx_{4.1\,1} \\ \dots \\ Finx_{4.1\,20} \end{array} \right) \\ Finx_{4.2} : \left( \begin{array}{c} Finx_{4.2\,1} \\ \dots \\ Finx_{4.2\,20} \end{array} \right) \end{array} \right] \end{array} \right\}$$

$$(3.3) \qquad A_{Max,\,Mary} = Finx_{1.1\,17}, Finx_{3.2\,15} \left| \int_{\text{Ц}2}^{\text{Ц}2} \right| = \{ Finx_{1.1\,17}, Finx_{3.2\,15} \}$$

Indeed, the formulation as a quantity is way easier to write, but it solely mentions the elements by themselves. If the elements are to be described in general, the integral formulation can be chosen.

### 3.3 Möbius Curvature

The shown fabric of MS appears to have different facets, and we have the ability to define new instruments of description within a well-defined fabric of the syntax. Still, the shown structure occurs to be static.

However, one of the requirements of 1.1.2 states that we must be able to formulate approximations. Approximations can be interpreted as the ability to cancel out different elements out of a situative matrix if we describe an event, but it also means that we may consider events as objects, or in terms of MS as finalities. Hence, there need to be a possibility to change elements' belonging to spaces.

In general, this also allows increasing the precision of description as well, since finalities can be then formulated as situative matrices, for instance.

As we change the precision grade of our description, every element occupies a different space (situant-situative-situatoric). This means, we slide down (or up) the precision grade, while the syntactic perspective stays the same. To explain it more figuratively, we may imagine the precision grade not to increase on an infinite, straight line, but one a Möbius strip: As we slide down (or up) the precision grade, situatoric elements become situatives, then situants (*vice versa*, respectively). As has been described in 3.1.2 and 3.1.3, both the situatoric and situant spaces bend into themselves. Hence, by changing the precision grade, the self-bending nature of the situatoric space transforms ultimately into the self-bending nature of situant space. If we interpret self-bending as a specific point on the Möbius strip, the two different spaces which occupy this point are on the opposite sides. By 'walking' far enough on the strip, one ends on the same point, but the opposite side. We may interpret this process as *Möbius curvature*. For application, we may use the symbol $M$:

$$(3.4) \qquad M^n_{X,\,Y}\left(E_{\text{Ц},\,1}\right) = E_{\text{Ц},\,2}$$

$n$ resembles the amount of minor space borders which have been crossed, and $X$ and $Y$ are the starting space and the finishing space respectively. The increasing of precision may be shown with a positive $n$, the decreasing of precision – approximations – with a negative $n$. The importance of the Möbius curvature will be shown in PART 2. PART 3 includes its explicit application.

## 4. Constructed Situations [11, 12, 13, 14, 19]

### 4.1 Construction as an Extension of Information

As has been shown in chapter two, relations are special situations which are part of the situative matrix in which the elements of relations are formulated. The reason for this formulation is provided by the comfortability of their notation as a tuple. Furthermore, this simple formulation prevents a redundant amount of meta information. Still, relations are situations according to the postulates, and we must treat them that way in the syntax. Thus, they are written within the situative matrix, separated by the situative and situant elements by straight lines to point out their situatoric nature, as well as other situators within the situation. In order to differentiate them from the matrical situations we initially formulate, they may be called *constructed situations*. The term 'constructed' illustrates that we, in fact, increase the amount of information by constructing links between the embedded elements. In contrast, all other elements may be called *basal situars* or *basal situaric elements*.

The symbol for constructed situations in general is *Konx*, whereas *x* represents the different types of constructed situations. In this chapter, we will first discuss relations concerning the spaces their elements belong to. Afterwards, we will introduce further constructed situations as more sophisticated instruments to increase the information by description. It is important to note, that these are solely suggestions. One can formulate other constructed situations or leave out the suggested ones, as long as the postulates are considered.

## 4.2 Classes of Relations

The *class* of a relation depends on its elements. If all elements are situatives, the relation is of a *situative* class. Accordingly, there are *situatoric* and *situant relations*. If the elements are part of different spaces, the according relations are called *situaric relations*. In general, relations are classified by the Greek letter $\rho$, thus the symbol for relations is *Konρ*. All classes and types of relations are summed up in **chart 4.1**.

### 4.2.1 Situative Relations

As has been shown in chapter two, there are three types of situative relations, $A\alpha$, $A\beta$ and $A\gamma$ and their mutated counterparts, $B\alpha$, $B\beta$ and $B\gamma$, respectively. They appear solely within situative matrices without crossing the edge of situative space and are the main type of situations to work with.

### 4.2.2 Situant Relations

If we observe a meta-element with a given amount of sub-elements, we can differentiate three types of relations as well. Actually, we would call a relation between a sub-element and the according meta-element situaric, since it crosses the border between situant and situative space. Due to comfortability, we may include this type into the situant class of relations. The symbol for this type may be *f*. If we formulate a relation between sub-elements of the same situant space, this is a situant relation of type *i*. As we have stated before, we must consider the situant symmetry, thus relations of type *i* include relations of

17

situants within the same column of a situative matrix as well. Finally, if we formulate a relation between situants of different situant spaces within a situative matrix, we have a relation of type *n*.

All situant relations are nominated by the letter *E*. After mutation, they are declared by the symbol *W*. Therefore, there are six possible situant relations: *Ef*, *Ei*, *En*, *Wf*, *Wi*, and *Wn*.

### 4.2.3 Situatoric Relations

When it comes to formulate relations between situators, we can differentiate three types as well. The most basal type is the relation between a situator and its meta-situation. This is a relation of type *a*. If a situator contains two sub-situators within its matrix, the relation between them is of type *b*. It is important to note that as long as we regard the first situator as our relevance, it does not matter whether the two other situators are both at the same 'sub-level' or not: One can contain the other or *vice versa*, or they may be both situatoric elements of the same sub-level with respect to the meta-situator. But as soon as we consider one of them as our relevance situation due to Möbius curvature, the *b*-relation mutates automatically into an *a*-relation, given the case that one situator is a part of the other. The third type, the *c*-relation, describes a relation between two situators without a known meta-situator. If one gains enough information to formulate a common meta-situator, this relation mutates into a *b*-relation. Non-mutated situatoric relations are nominated by the Greek letter *Γ* and mutated relations by the Greek letter *Δ*, respectively.

### 4.2.4 Situaric Relations

Situaric relations have a much greater range of types to occur in, since there are more possibilities to cross spaces than to stay within their ranges. First, the *π-relation* describes the basal relation between a situative and its matrical situation. Type *ρ* relations occur between a situative and a situator within the same situative matrix. The *σ-relation* describes a relation between a situative and another situator, which is not a sub-situator of the situative's situator, whereas the two situations share a common meta-situator. Relations of type *τ* resemble the same case, whereas there is no common meta-situator. Type *ϕ* and *χ* relations, finally, contain situative, situant elements, respectively as well as elements of any other space, except their own meta-element.

The reason why we need to distinguish relations between situatives and situators with a high precision is that the situative matrix and the situatoric space are the common range of analysis. Relations with situants can be summed up into one relation, since we simply can apply Möbius curvature if we intend to investigate in this part with higher precision. Otherwise, if we would distinguish all different combinations, the amount of types would exceed into redundancy. Situaric relations are nominated by the Greek letter *Φ*, and by *Ω* after mutation, respectively.

**Chart 4.1:** Summary of relations and their mutated counterparts with their corresponding spaces and element symbols which are involved in the relations, as well as numbers of possible types. Basal means non-mutated.

| Space (name, symbol) | Element (symbol) | Relations (basal, mutated) | | Types (basic to complex) |
|---|---|---|---|---|
| Situaric, Ц | $E_Ц$ | Φ | Ω | π ρ σ τ φ χ |
| Situatoric, C | $E_C$ | Γ | Δ | a b c |
| Situative, З | $E_З$ | A | B | α β γ |
| Situant, Ч | $E_Ч$ | E | W | f i n |

The classes and types of relations are, as all constructed situations, solely a suggestion. The amount of classes or types can be changed if it is useful for simplification of the syntax or for efficiency maximization.

## 4.3 Operators

As has been said in 3.3, the MS syntax is still quite static. The first attempt to make the syntax more dynamical was the introduction of the Möbius curvature in 3.4. One can interpret the Möbius curvature as a sort of function, which assigns one situar element to another. Since we have stated that we need to distinguish basal situar elements from constructed situations, it is helpful to distinguish the assignment process accordingly. I have chosen to call assignments with solely basal situar elements as *operators* and leave the term *function* for assignments in which constructed situations occur.

### 4.3.1 Operators as Connections between Basal Situar Elements

If situations change, they lose their integrity and thus become another situation. This means, that the concept of time does not exist per se within the syntax. Since we need to introduce dynamical elements for proper description, we need to connect situations with each other to obtain a chain of static situations, which then results in a moving picture, similar to the mode of operation of a movie.

Thus, we introduce the *operator Konω* as a constructed situation which assigns one situation to another, as we have demanded before. Operators can not solely introduce an arrow of time into the syntax, but can also describe generally deterministic or even random evolution. This general approach is crucial in order to satisfy the requirements stated in 1.1.2.

In order to set up both formulation and possibilities of operators, causality and logical operators will be discussed, as well as the random operator as an example for non-deterministic assignment.

### 4.3.2 Causality and Logical Operators

Since we are used to observing events in the subabstract sphere, it is important to discuss the *causality operator* first. It assigns one situar to another in a determined way, whereas both situars can be distinguished from each other by space-time coordinates (and other attributes). In fact, we can choose whether we go in the direction of the experienced flow of time or not, since both directions are important for application of scientific theories in terms of predictions and calculations.

The causality operator can be symbolized by the letter $K$, whereas the known steps are written in an index on the left side and the minor spaces of the according elements on the right side. If both elements are of the same minor space, we can write solely that minor space as the right index. An application example is shown by (4.1).

$$
(4.1) \qquad Sit\sigma_1 := \left\{ \left. \begin{matrix} Sit\varepsilon_1 \\ Sit\eta_1 \;\; Sit\eta_2 \\ Finx_1 \;\; Finx_2 \end{matrix} \right| \left. \begin{matrix} Kon\rho_{1,2} \end{matrix} \right| \right\}
$$

with

$$
Kon\rho_{1,2} = A\overset{2}{\gamma}\begin{pmatrix} 0 + 0 \\ 1 \;\; - \;\; - \end{pmatrix} < Finx_1 \,; Sit\varepsilon_1 \,; Sit\eta_2 \,; Finx_2 >
$$

Furthermore

$$
Sit\sigma_2 := \left\{ \begin{matrix} Sit\varepsilon_1 \\ Sit\eta_1 \;\; Sit\eta_2{}' \\ Finx_1 \;\; Finx_2 \end{matrix} \right\} \qquad Sit\sigma_3 := \left\{ \begin{matrix} Sit\varepsilon_1 \\ Sit\eta_1 \;\; Sit\eta_2{}' \\ Finx_1 \;\; Finx_2{}' \end{matrix} \right\}
$$

with

$$
Finx_2 \neq Finx_2{}' \qquad Sit\eta_2 \neq Sit\eta_2{}'
$$

20

Now we can formulate two causality operators between the three situations:

(4.2) $$_1K_C\left(Sit\sigma_1\right) = Sit\sigma_2 \qquad _1K_C\left(Sit\sigma_2\right) = Sit\sigma_3$$

In order to formulate the explicit change from one to another element, we can write the operator as a situator with a situative matrix. Within this matrix, we can place both elements we want to connect and add as much other elements as we want, as well as new construct situations which show the connection between the elements.

Causality occurs solely in the subabstract sphere. But deterministic connections also appear between elements of the abstract sphere. According to the sphere model depicted in 1.2.1, determinism is in fact implied by the abstract sphere. For instance, numbers can be connected by the Fibonacci sequence, which is deterministic by definition. Subabstract manifestations of this sequence are found in several patterns of subabstract elements, i.e. in botanic. Therefore, it is helpful to introduce an operator which solely connects elements outside the subabstract sphere in order to differentiate these types of situaric elements. The deterministic operator for abstract elements may be called the *logical operator F*.

Both types of operators can not only show a deterministic connection between two events, but also show chains of connections between arbitrary amounts of situar elements. In order to illustrate this property, we introduce the *operatoric viality*. Similar to the relation viality we discussed in 2.3.2., operatoric viality shows the operator's path throughout the situar space. It can be seen as a chain of single operators and is written within an index in the right upper corner with respect to the operator symbol. We can either simply declare the amount of situars between the argument and the final situar (object situars of type two, according to 2.3.1) or write them down in parentheses (4.2).

(4.2) $$_1F_C^{2\left(Sit\sigma_2,\ Sit\sigma_3\right)}\left(Sit\sigma_1\right) = Sit\sigma_4$$

### 4.3.3 Coincidence Operator

Whenever we do not know why an outcome of an event came to be or if there is no certainty of which successor an event will have (for instance, in quantum mechanics), we describe this as a coincidence. In fact, these are two different types of uncertainty, but initially, we can formulate both types in the same way. Since there is no clear reason, whether subjectively or objectively, why the observed successor occurred, we cannot formulate the connection with a deterministic operator. Thus, we introduce the *coincidence operator Z*.

If formulated as a situator, there are no deterministic formulations between the two assigned situars within the situative matrix. Still, we can show how many possible outcomes exist, or at least how many we know. In order to include this information into the notation, we write the amount of the possible elements right above the operator symbol, as (4.3) shows.

(4.3)
$$\overset{2\ \ 1}{\underset{1\ \ C}{Z}}\left(Sit\sigma_1\right) = Sit\sigma_2,\ Sit\sigma_3$$

In general, we can introduce any possible operator we wish. Even nonsense operations are allowed according to the postulates. The Möbius curvature can be formulated as an operator too. Henceforth, we will refer to it as the $M$ operator. In (3.4), we already used this notation.

## 4.4 Functions

As has been stated in 4.3., operators are situations which connect basal situar elements. Whenever at least one of the elements is a constructed situation, we may call that connection a *function*. The formalism for a function may be $Kon\phi$. Consequently, functions in terms of MS are also able to assign operators to each other, which is indeed unfamiliar with respect to the function-operator relation in mathematics. However, this avoids confusion between these two languages.

There may be causality, logical and coincidence functions, as well as any kind one can use for any given purpose. For instance, we can sum up operators by functional assignment. In the situative matrix shown in (4.1), we can formulate two different coincidence operators, one of them being written as:

(4.4)
$$Kon\varpi_2 = {}_1Z^1_C\left(Sit\sigma_1\right) = Sit\sigma_3 \in Sit\sigma_M$$

Both operators have solely one outcome and can be written as a part of the situation shown in (4.5). In the next step, we apply a function which combines these operators to provide a third one, as shown in (4.6). The definition of both instruments is intended to provide unrestricted possibilities for application and combination of operators and functions.

(4.5)
$$Sit\sigma_M := \left\{ \left\| \begin{matrix} Sit\sigma_1 & Kon\varpi_1 & Kon\phi_{1,\,2,\,3} \\ Sit\sigma_2 & Kon\varpi_2 & \\ Sit\sigma_3 & Kon\varpi_3 & \end{matrix} \right\| \right\}$$

(4.6)
$$Kon\phi_{1,\,2,\,3}\left(Kon\varpi_1, Kon\varpi_2\right) = Kon\varpi_3 = {}_1\overset{2\ \ 1}{Z_C}\left(Sit\sigma_1\right) = Sit\sigma_2, Sit\sigma_3$$

Since functions are constructed situations, they can, according to their definition, also be arguments of other functions as well. An example is given by the situative matrix in (4.7), in which a general transition of information between the finalities is given, represented by the relation (4.8).

$$
(4.7) \qquad Sit\sigma_R := \left\{ \begin{matrix} & & & & & \begin{vmatrix} Kon\rho_{1,2} & Kon\rho_{4,5} & Kon\phi_M \\ Sit\eta_1 & Sit\eta_2 & Sit\eta_3 & Sit\eta_4 & Sit\eta_5 & Kon\rho_{2,3} & Kon\phi_{1,2,2,3} & Kon\phi_{1,2,4,5} \\ Finx_1 & Finx_2 & Finx_3 & Finx_4 & Finx_5 & Kon\rho_{3,4} & Kon\phi_{3,4,4,5} \end{vmatrix} \end{matrix} \right\}
$$

$$
(4.8) \qquad Kon\rho_{a,b} = \begin{pmatrix} + \\ 0 \end{pmatrix} < Finx_a \,;\, Finx_b > \quad a,b \in \mathbb{N}
$$

Now, we can formulate a function between $Kon\rho_{b,c}$ and $Kon\rho_{a,b}$.

$$
(4.9) \qquad Kon\phi_{a,b,b,c}\big( Kon\rho_{a,b} \big) = Kon\rho_{b,c} \quad a,b,c \in \mathbb{N}
$$

Since we solely observe five finalities within the situation, we can cross the combinations between $Finx_1$ and $Finx_3$ as well as between $Finx_3$ and $Finx_5$ to show transitivity of the depicted relations:

$$
(4.10) \qquad Kon\phi_M\big( Kon\phi_{1,2,2,3}, Kon\phi_{3,4,4,5} \big) = Kon\phi_{1,2,4,5}
$$

In general, functions are not bound to situative matrices. This means, they can be applied within the whole situaric space, independent of minor spaces. Space intersecting functions are indeed quite important for conclusions and pattern recognition, which makes them a crucial tool for research processes. A simple example is demonstrated by the system (4.11).

$$
(4.11) \qquad Sit\sigma_1 := \left\{ \begin{matrix} & \begin{vmatrix} Kon\rho_{1,1} & Kon\phi_{1,2} \\ Sit\eta_1 & Kon\rho_{1,2} \\ Finx_1 & Finx_2 & Kon\rho_{2,1} \end{vmatrix} \end{matrix} \right\}
$$

and

$$
Sit\sigma_2 := \left\{ \begin{matrix} Sit\varepsilon_3 & \begin{vmatrix} Kon\rho_{3,3} & Kon\phi_{3,4} \\ Sit\eta_3 & Kon\rho_{3,4} \\ Finx_3 & Finx_4 & Kon\rho_{4,3} \end{vmatrix} \end{matrix} \right\}
$$

with

$$
Sit\sigma_1, Sit\sigma_2 \in Sit\sigma_M := \left\{ \begin{matrix} \begin{Vmatrix} Sit\sigma_1 & Kon\phi_{1,2,3,4} \\ Sit\sigma_2 \\ Kon\rho_{1,2,3,4} \end{Vmatrix} \end{matrix} \right\}
$$

23

The sets of relations and functions can be described in the same manner as above:

$$(4.12) \qquad Kon\rho_{a,a} = \begin{pmatrix} 0 \\ + \end{pmatrix} < Finx_a \, ; Sit\,\eta_a > \quad a \in \mathbb{N}$$

$$Kon\rho_{b,a} = \begin{pmatrix} - \\ + \end{pmatrix} < Finx_b \, ; Sit\,\eta_a > \quad a,b \in \mathbb{N}, a < b$$

$$Kon\rho_{a,b} = \begin{pmatrix} + \\ 0 \end{pmatrix} < Finx_a \, ; Finx_b > \quad a,b \in \mathbb{N}, a < b$$

$$Kon\phi_{a,b}\left(Kon\rho_{a,a}, Kon\rho_{b,a}\right) = Kon\rho_{a,b} \quad a,b \in \mathbb{N}, a < b$$

Given the relation pattern, we can now apply a space intersecting function which provides a $\Phi\phi$-relation to the known functions. This new relation represents the conclusion that both sets of finalities behave similarly.

$$(4.13) \qquad Kon\phi_{1,2,3,4}\left(Kon\phi_{1,2}, Kon\phi_{3,4}\right) = Kon\rho_{1,2,3,4} = \Phi\phi < Finx_1; Finx_2 \, ; Finx_3 \, ; Finx_4 >$$

## 4.5 Perspectives

After the introduced instruments have transformed the static MS syntax into a more dynamical one, one last constructed situation will be discussed now. A proper formulation of perspectives is crucial for third person view of information gain and formulation of conditions for self-awareness, thus for development of artificial intelligence as well. The required type of constructed situations is called *perspectives* or $Kon\pi$. A perspective is defined by the elements the according finality, called *perspector*, can observe. They can be included in a summation of space parts, representing the range of the perspective. This range is called *copula* $\mathcal{K}$. An example is shown by (4.14).

$$(4.14) \qquad A_1, A_2, ... \in \mathcal{K}$$

### 4.5.1 Different Types of Perspectives

If we intend to illustrate the different ranges of perspectives, we need to declare the differences of their according copulae. Dependent on which space parts within a given space are included by the copula, like the situative, we can differentiate between two possible kinds of perspectives. (4.15) shows an example which clarifies the difference by pointing out the colors three persons ($Finx_1$, $Finx_2$ $Finx_3$) can see. The perspective is symbolized by a cursive '$P$'.

24

$$(4.15) \qquad Sit\sigma_1 := \left\{ \begin{array}{ccc|c} & & & \begin{array}{cc} Sit\sigma_o & Kon\pi_3 \end{array} \\ Sit\eta_1 & Sit\eta_2 & Sit\eta_3 & Kon\pi_1 \\ Finx_1 & Finx_2 & Finx_3 & Kon\pi_2 \end{array} \right\}$$

with

$$Sit\sigma_o := \left\{ \begin{array}{ccc} Sit\eta_{red} & Sit\eta_{green} & Sit\eta_{other\ colors} \\ Finx_{red} & Finx_{green} & Finx_{other\ colors} \end{array} \right\}$$

with two perspectives of the first finalitites:

$$Kon\pi_1 = P : A_1 = E_{\mathcal{U}} \left| \int_3^C \right| = \left\{ Sit\sigma_o, Finx_{green}, Sit\eta_{green}, Finx_{other\ colors}, Sit\eta_{other\ colors} \right\} \in \mathcal{K}_1$$

$$Kon\pi_2 = P : A_2 = E_{\mathcal{U}} \left| \int_3^C \right| = \left\{ Sit\sigma_o, Finx_{red}, Sit\eta_{red}, Finx_{other\ colors}, Sit\eta_{other\ colors} \right\} \in \mathcal{K}_2$$

The first person cannot see red color, whereas the second one is unable to see green color. The third person has no visual impairment and is able to detect all colors within the situative matrix of $Sit\sigma_0$. The difference between $Finx_1$ and $Finx_2$ on the one hand and $Finx_3$ on the other hand is the amount of elements engulfed by their copulae. While the copulae of $Finx_1$ and $Finx_2$ are defined by space parts of $Sit\sigma_0$, the copula of $Finx_3$ is defined by the whole situative space of the matrix. The first two perspectives may be called *relative*, the third one *absolute*. Accordingly, the perspectives are written with an 'r' or an 'a' as an index, respectively.

### 4.5.2 Promotion and Interlation Functions

In the next step, we need to clarify the connection between relative and absolute perspectives. The formulation of this development becomes crucial when it comes to analysis processes, in which information is gained and the copula of a relative perspective increases. This leads ultimately to the formation of an absolute perspective, as soon as the copula engulfs a whole minor space. That development can be formulated by a function applied on the relative perspective, which provides an absolute perspective in the outcome, as is shown in (4.16).

$$(4.16) \qquad \boldsymbol{P}\left( Kon\pi_i = P_r \right) = Kon\pi_j = P_a \qquad i \neq j$$

25

The function is called *promotion function* **P** and can be applied backwards to represent a decline of the described copula, i.e. as a declaration of information loss. Similarly to the example in (4.10), we can also build a chain of promotion functions, which provide ever-increasing copulae. To sum up the interlation of copulae between each other, we introduce the *interlation function E*. Instead of perspectives, this function is applied on copulae. The promotion function can be applied on copulae as well, but the perspectives themselves have been chosen in order to declare the change from relative to absolute type. However, one can change this suggestion if it is helpful. Generally, the interlation function can be written like shown by (4.17).

(4.17) $$E^q(\mathcal{K}_a) = \mathcal{K}_b \qquad q \in \mathbb{Q} \ \ a, b \in \mathbb{N}, a \neq b$$

If the interlation process is repeated endlessly, the copulae of absolute perspectives become ever bigger. As has been stated in the postulates, there are infinitely many situations. It is yet unclear whether an interlation progression would be divergent or convergent and has to be shown in further research. Nevertheless, we can formulate at least a hypothetical perspective, the *global perspective*, defined by a copula which is the limes of an interlation progression applied on an arbitrary copula, as (4.18) shows.

(4.18) $$\mathcal{K}_g = \lim_{q \to \infty}\left(E^q(\mathcal{K}_a)\right)$$

### 4.5.3 Paradox Descriptions

As discussed in 1.1.2, one of the requirements of an extraabstract language is the ability to describe paradoxes. Whenever a paradox occurs, one can distinguish two perspectives of a situation, which contradict each other. By formulating the different copulae of all included perspectives, one can deduce the situative matrix with its other constructed situations and, by combination, formulate the paradox as a whole. Since each perspective can be seen as a situation by itself, both perspectives describe a consistent system each. Therefore, the absolute perspective over both copulae describes the paradox situation as a whole. Whether and how the paradox nature can be dismissed by this process in general, has to be shown by further research.

## 5. Attributes [12, 13, 20]

In the previous chapters we have introduced the basal elements of the MS syntax and suggested additional instruments for precise description. Now, attributes and their accoridng affects will be discussed. As information is stored in attributes, they will be discussed first. Chapter five will give an insight on the nature of affects.

In order not to over expand the content of this work, solely the basic analysis of attributes will be presented, and accordingly, the discussion of affects will solely proceed as far as it is necessary to formulate proper research.

## 5.1 Basal Similiarity Grade

Recognition of patterns is crucial for research, as has been discussed in 4.4. Accordingly, we must be able to describe similarity of finalities. Attributes in general can be formulated as meta-elments to other attributes, whereas their sub-elements are solely those we know about in the description. The relation between the quantities of sub-attributes $E_A$ and $E_B$ of two finalities $Fin_A$ and $Fin_B$ is the *basal similarity grade* $Æ_{A,B}$. It is defined by the formula given in (5.2).

$$(5.1) \qquad Sit\sigma_1 := \left\{ \begin{array}{ll} Sit\,\eta_A : \begin{pmatrix} Sit\,\eta_{A1} \\ ... \\ Sit\,\eta_{Ap} \end{pmatrix} & Sit\,\eta_B : \begin{pmatrix} Sit\,\eta_{B1} \\ ... \\ Sit\,\eta_{Bq} \end{pmatrix} \\ Finx_A : \begin{pmatrix} Finx_{A1} \\ ... \\ Finx_{Ap} \end{pmatrix} & Finx_B : \begin{pmatrix} Finx_{B1} \\ ... \\ Finx_{Bq} \end{pmatrix} \end{array} \right\}$$

with

$$E_A = \left\{ Sit\,\eta_{A1}, ..., Sit\,\eta_{Ap} \right\} \qquad E_B = \left\{ Sit\,\eta_{B1}, ..., Sit\,\eta_{Bq} \right\}$$

$$(5.2) \qquad Æ_{A,B} = \frac{|E_A \cap E_B|}{|E_A| + |E_B|}$$

Its value can differ between one and zero. If $Æ_{A,B}=0$, both attributes can be regarded as completely different. If $Æ_{A,B}=1$, one can say they are qual to each other, at least given the amount of information provided by (5.1). This seems intuitive, since the highest grade of similarity implies equivalence. All values in between indicate that both attributes share common sub-attributes, which makes them more or less similar to each other.

The term ‚basal‘ means that we solely include the different sub-elements in our calculation, but not the combination of them. In fact, attributes can cancel out each other, or even produce new attributes, which will be discussed later in this chapter.

As has been shown in the previous chapter, we can formulate transitivity by applying functions on relations between finalities. We can interpret (5.2) as a relation as well. Thus, we can apply this quantitative characterization to discuss the concept of transitivity which has been mentioned in 4.4. (5.3) shows a situative matrix with three finalities; each finality has its amount of sub-elementar attributes.

$$(5.3) \qquad Sit\sigma_1 := \left\{ \begin{array}{cccc} Sit\,\eta_A: \begin{pmatrix} Sit\,\eta_{A\,1} \\ ... \\ Sit\,\eta_{A\,p} \end{pmatrix} & Sit\,\eta_B: \begin{pmatrix} Sit\,\eta_{B\,1} \\ ... \\ Sit\,\eta_{B\,q} \end{pmatrix} & Sit\,\eta_C: \begin{pmatrix} Sit\,\eta_{C\,1} \\ ... \\ Sit\,\eta_{C\,r} \end{pmatrix} \\ Finx_A: \begin{pmatrix} Finx_{A\,1} \\ ... \\ Finx_{A\,p} \end{pmatrix} & Finx_B: \begin{pmatrix} Finx_{B\,1} \\ ... \\ Finx_{B\,q} \end{pmatrix} & Finx_C: \begin{pmatrix} Finx_{C\,1} \\ ... \\ Finx_{C\,r} \end{pmatrix} \end{array} \right\}$$

Furthermore, let be $Æ_{A,B}=l$ and $Æ_{B,C}=m$. Then we can differentiate between three cases:

1) If $l+m=2$, $l=m=1$ and all three finalities are equal to each other. Accordingly, $Æ_{A,C}=1$.

2) If $1<l+m<2$, the amounts of common situant attributes of both pairs must have elements which occur in the corresponding other. Hence, the intersection amount is greater than zero: $0< Æ_{A,C}<1$.

3) If $0\leq l+m\leq 1$, there is a probability unequal to zero that the same situant attributes occur in both pairs, whereas there may be none common attributes between $Fin_A$ and $Fin_B$ at all, too. Thus, $0\leq Æ_{A,C} \leq 1$. The probability for $Æ_{A,C}$ being unequal zero can be calculated:

$$(5.4) \qquad P(Æ_{A,C}) = \frac{Æ_{A,B} \cdot Æ_{B,C}}{Æ_{A,B} + Æ_{B,C}}$$

The discussed cases represent different transitivity grades. The first case is called *total transitivity*:

$$(5.5) \qquad Æ_{A,B} + Æ_{B,C} = 2 \;\Rightarrow\; Æ_{A,B} = Æ_{B,C} = Æ_{A,C} = 1$$

The second case describes a *partial transitivity*:

$$(5.6) \qquad Æ_{A,B} + Æ_{B,C} \in \,]1;2[ \;\Rightarrow\; Æ_{A,C} \in \,]0;1[$$

The latter case describes what is called *latent transitivity*:

$$(5.7) \qquad Æ_{A,B} + Æ_{B,C} \in [0;\,1[ \;\Rightarrow\; Æ_{A,C} \in [0;\,1[$$

## 5.2 Attribute Combinatorics

In order to differentiate between attributes according to their behavior, it is a useful approach to analyze their behavior in combination, as mentioned in 5.1. Thus, the different types of combinations need to be discussed.

### 5.2.1 Types of Combinations

Assume a given number $n$ of attributes. If they are combined to give a meta-attribute, its composition may differ from the initial set of attributes. Let such a situative matrix $Sit\sigma_2$ may be the outcome of an applied operator on the initial set of attributes $Sit\sigma_1$:

$$(5.8) \qquad Sit\sigma_1 := \left\{ Sit\eta_1, \ldots, Sit\eta_n \right\} \quad n \in \mathbb{N}$$

$$Sit\sigma_2 := \left\{ Sit\eta_A : \begin{pmatrix} Sit\eta_{A1} \\ \ldots \\ Sit\eta_{Ar} \end{pmatrix} \right\} \quad r \in \mathbb{N}$$

Let there further be:

$$(5.9) \qquad m, l, o \in \mathbb{N}, \quad m < n, \; l \leq n$$

The given numbers do not solely represents the amount of attributes, but also different sets. If the total number of attributes remains while the composition has been changed, the resulting variable changes as well. Given these frame conditions, we can distinguish six different types of combinations.

### 5.2.1.1 Omnitive Combinations

In the simplest case, all different attributes remain unchanged after combination to a meta-attribute. This case occurs when we solely observe properties like mass. For instance, we may measure the mass of a set of parts of a building. After the building is completed, the mass simply adds up, whereas nothing of it gets lost. Such combinations are called *omnitive*, whereas $r=n$.

### 5.2.1.2 Complutive Combinations

In the second case, not all properties remain after combination. Given the property of mass, this occurs under relativistic conditions, for example in element synthesis in the center of stars. A part of the mass is transferred into energy, thus it gets lost, given solely mass is observed. Those combinations are called *complutive*, whereas $r=m$.

### 5.2.1.3 Monotive Combinations

A special type of complutive combinations is represented by *monotive combinations*. They occur if solely one of all initial attributes remain after combination. Given the example in 5.2.1.2, a collapsing star can be transformed into a black hole if it is sufficiently massive. While the attribute of mass (approximately) remains, other observed attributes such as shape, luminosity and temperature vanish. In this case, $r=1$.

### 5.2.1.4 Nihilitive Combinations

A far extreme case of combinations occurs, when all observed attributes vanish after combination. If we observe mass of equal amounts of matter and antimatter, it vanishes completely after combination of both. In this case, the combination is *nihilitive*. Accordingly, $r=0$.

### 5.2.1.5 Transitive Combinations

Attributes cannot solely be canceled out, but can also emerge in the combination process. If we observe not only mass but also energy in the case of 5.2.1.4, the given mass vanishes completely in order to be replaced by energy, according to the well-known principle of equivalence of mass and energy. One can say, the initial set of attributes has been transformed into another one. This type of combinations may be called *transitive*. Since it provides a new, different set of properties, $r=o$.

### 5.2.1.6 Additive Combinations

Finally, if none of the given attributes are destroyed, whereas a given quantity, $l$, is created, the combination is *additive*. This may occur if atoms with observed mass are combined to obtain molecules. Their masses add up to give the molecule's mass. The new properties, such as color or conductivity, are created by combination, though. In this case, $r=l+o$.

## 5.2.2 Combinatorial Types of Attributes

According to the presented types of combinations, we can deduce which combinatorial types of attributes there are. They can be depicted if we sub classify the initial set in two groups of attributes, which have a meta-attribute each. There are three classes of meta-attributes which, by combination, lead to the six possible combination types as shown in 5.2.1.:

1) Attributes can create new ones if combined. This implies transitive or additive combinations. These are *novial attributes*.

2) Comlutive, monotive, transitive and nihilitive combinations are implied by attributes which annihilate each other if combined. These are *annihilating attributes*.

3) If the initial set of attributes remains unchanged, an omnitive combination is observed, which can solely occur if there are *neutral attributes*.

Novial and annihilating attributes can be differentiated more precisely, which results in six combinatorial types of attributes.

If the initial sets of attributes decrease, whereas new ones are created as well, the attributes are *partially additively novial*. The declaration is $Sit\eta^{pan}$. Another case occurs if the combination is additive. The initial sets are extended by new attributes. This is caused by *totally additive novial* attributes. They are declared by $Sit\eta^{tan}$. If the combination is transitive, the initial sets of attributes vanish completely and new ones are created. This case occurs when the combined attributes are *transitively novial*. They are written as $Sit\eta^{trn}$. Annihilating attributes can behave in two ways. Either they vanish all the counterpart's elements, or solely a given number of its set. I the first case, we they are *totally annihilating*, in the second case *partially annihilating*. Their declarations are $Sit\eta^{tn}$ and $Sit\eta^{pn}$, respectively.

## 5.3 Summary of Attribute Combinations

If two attributes of different types are combined, the type of combination may differ. In order to clarify which type of combination occurs in this case, we need to formulate the affect on the corresponding other attribute and combine them. In order to keep the formulation simple, we differentiate between three types of each attributes' sets: The first type is the initial set $m$. If some attributes of the initial set are vanished, the reduced initial set is called $k$. Every created attribute can be summed up in the set $o$.

The meta-attributes are called $Sit\eta_A$ and $Sit\eta_B$. $Sit\eta_A$ implies the combinatorial behavior of $Sit\eta_B$ and *vice versa*.

Attributes of type *tan* imply the set *m+o* for the corresponding attribute. Attributes of type *pan* imply the set *k+o*, whereas type *trn* attributes imply the set *o*. Type *tn* attributes imply a set equal zero, and type *pn* attributes imply the set *k*. Finally, neutral attributes do not have any affect on the corresponding attribute, thus they leave the corresponding set unchanged. Sets implied by $Sit\eta_A$ are written with index $\beta$, and *vice versa*. The new set $r_{A,B}$ of the combined attributes is expressed by the sum of both $\alpha$ and $\beta$ sets. **Chart 5.1** shows the different results of combinations of all six types of attributes.

**Chart 5.1:** Calculation of sets of combined attributes $Sit\eta_A$ and $Sit\eta_B$. m=initial set, k=reduced initial set, o=additional set. The indices $\alpha$ and $\beta$ represent the belonging to $Sit\eta_A$ and $Sit\eta_B$ respectively. All sets of each attribute are summed up as $r_x$, $x = \alpha$, $\beta$. $r_{A,B} = r_\alpha + r_\beta$ is the set of combined attributes.

| $r_{A,B}$ $Sit\eta_B$ | $Sit\eta_A$ $r_x$ | tan $m_\beta + o_\beta$ | pan $k_\beta + o_\beta$ | trn $o_\beta$ | tn $0$ | pn $k_\beta$ | n $m_\beta$ |
|---|---|---|---|---|---|---|---|
| **tan** | $m_\alpha + o_\alpha$ | $m_\alpha + m_\beta + o_\alpha + o_\beta$ | $m_\alpha + k_\beta + o_\alpha + o_\beta$ | $m_\alpha + o_\alpha + o_\beta$ | $m_\alpha + o_\alpha$ | $m_\alpha + o_\alpha + k_\beta$ | $m_\alpha + m_\beta + o_\alpha$ |
| **pan** | $k_\alpha + o_\alpha$ | $m_\beta + k_\alpha + o_\alpha + o_\beta$ | $k_\alpha + k_\beta + o_\alpha + o_\beta$ | $k_\alpha + o_\alpha + o_\beta$ | $k_\alpha + o_\alpha$ | $k_\alpha + k_\beta + o_\alpha$ | $m_\beta + k_\alpha + o_\alpha$ |
| **trn** | $o_\alpha$ | $m_\beta + o_\alpha + o_\beta$ | $k_\beta + o_\alpha + o_\beta$ | $o_\alpha + o_\beta$ | $o_\alpha$ | $k_\beta + o_\alpha$ | $m_\beta + o_\alpha$ |
| **tn** | $0$ | $m_\beta + o_\beta$ | $k_\beta + o_\beta$ | $o_\beta$ | $0$ | $k_\beta$ | $m_\beta$ |
| **pn** | $k_\alpha$ | $m_\beta + k_\alpha + o_\beta$ | $k_\alpha + k_\beta + o_\beta$ | $k_\alpha + o_\beta$ | $k_\alpha$ | $k_\alpha + k_\beta$ | $m_\beta + k_\alpha$ |
| **n** | $m_\alpha$ | $m_\alpha + m_\beta + o_\beta$ | $m_\alpha + k_\beta + o_\beta$ | $m_\alpha + o_\beta$ | $m_\alpha$ | $m_\alpha + k_\beta$ | $m_\alpha + m_\beta$ |

This chart has an important role for MS syntax and will be used in chapter six to elaborate a detailed formulation of affects. Indeed, there are further researches required, especially when more than two attributes are combined. The discussion of such cases would extend this work redundantly, as its aim is the discussion of approaches for precise descriptions.

# 6. Affects [8, 12, 13, 20]

## 6.1 Connection between Attributes and Affects

As has been stated in 3.1.3, situant elements within a column underlie the principle of situant symmetry. This principle implies that if the intrinsic integrity of attributes is changed by combinaitons, the same change occurs for the corresponding affects as well. Hence, the combinatorial statements evolved in chapter five can be applied in the same way on affects. However, important affect vocabulary needs to be introduced first.

## 6.2 Compatibility of Attributes and Affects

### 6.2.1 Derivation

Similarly to affects depending on their according attributes, the objects' attributes also imply by which affects the corresponding finalities can be affected. Thus, both attributes have to be included for a proper description. In order to determine whether and in which extent a finality affects another, the *compatibility C* is introduced as a function of subject and object attributes and the observed affect itself. Alternatively, $C$ can be formulated as a function of the basal similarity grade we introduced in 5.1, since a finality can only have an affect on finalities with attributes similar to those which contribute to the observed affects.

$$(6.1) \qquad C\left(Sit\varepsilon_{A,} Finx_B, Sit\eta_B\right) = f\left(\cancel{E}_{A,B}\right)$$

An affect occurs, if it depicts finalities with attributes which allow it to affect the finality. Besides the basal similarity, this has to be considered as well. To formulate the relevance of attribute for an affect in a quantitative way, the r*elevance factor r* is introduced. $r$ is a function of both the affect itself (or the subject's attribute, alternatively) and the object's attribute. How the relevance factor is determined explicitly, will be discussed in 6.5.1. Given both the basal similarity and the relevance factor, the compatibility can be formulated as:

$$(6.2) \qquad C\left(Sit\varepsilon_A, Finx_B, Sit\eta_B\right) = \cancel{E}_{A,B} \cdot \sum_{i=0}^{\left|Sit\eta_B\right|} r\left(Sit\eta_{Bi}\right)$$

### 6.2.2 Distinction of Attributes according to Compatibility

We can differentiate three types of attributes with respect to their Compatibility to a given affect:

1) *Adaptive attributes* are compatible with the affect and change afterwards according to the affect-implying attributes.

2) *Affine attributes* are compatible with the affect, but do not change after the affection.

3) *Inert attributes* are not compatible with the affect.

It is important to note, that 2) and 3) describe two different compatibility types of neutral attributes (5.2.2).

## 6.3 Copula Potential

In 4.5, we introduced the copula as the sum of all space parts the perspector is able to detect. If we want to derive an explicit formulation for the potential of the copula, that is the potential to be aware of things, we need to examine the basic process of detection.

If a finality $Fin_A$ detects $Fin_B$, we can interpret this process as an exchange of affects which enable detection. On the one hand, $Fin_B$ can only be detected by $Fin_A$ if it is *detectable* by $Fin_A$, which can be formulated by the compatibility introduced in 6.2.1. On the other hand, $Fin_A$ can only detect $Fin_B$ if the affects originating in $Fin_B$ are compatible with the attributes of $Fin_A$. The interpretation of detection as an interdependence helps to find an approach to the derivation of the copula potential. However, it is important to note that not only visual detection must be considered.

In order to obtain a quantitative formulation, we focus on two sets of attributes:

$$(6.3) \qquad Sit\,\eta_A : \begin{pmatrix} Sit\,\eta_{A1} \\ ... \\ Sit\,\eta_{A\alpha} \end{pmatrix}, \quad Sit\,\eta_B : \begin{pmatrix} Sit\,\eta_{B1} \\ ... \\ Sit\,\eta_{B\beta} \end{pmatrix}$$

Since we need to include the relevance factor for each situant attribute of $Sit\,\eta_B$ and for each situant attribtue of $Sit\,\eta_A$ as well, the compatibility is written as:

$$(6.4) \qquad C\left(Sit\varepsilon_A, Finx_B, Sit\eta_B\right) = C_{A,B} = \text{\AE}_{A,B} \cdot \sum_{j=1}^{\alpha} \sum_{i=1}^{\beta} r_{Aj}\left(Sit\eta_{Bi}\right)$$

To shorten the formulation, we introduce a new term (6.5) for the double sum. Now, we can reformulate (6.4) into (6.6), whereas both ways have been considered (*A-B, B-A*).

$$(6.5) \qquad S_{rA}^{\alpha\beta}\left(Sit\,\eta_A, Sit\,\eta_B\right)$$

$$(6.6) \qquad C_{A,B} = \text{\AE}_{A,B} \cdot S_{rA}^{\alpha\beta}\left(Sit\,\eta_A, Sit\,\eta_B\right) \qquad C_{B,A} = \text{\AE}_{B,A} \cdot S_{rB}^{\beta\alpha}\left(Sit\,\eta_B, Sit\,\eta_A\right)$$

34

In the next step, we can formulate the copula potential $\text{Đ}$ between both finalities as a product of both compatibilities, since they both represent the affection between them. It is important to note that the potential of the copula is not the copula itself. However, by finding the maximum copula potential, one can derive that the according attributes lead to an absolute perspective within a given minor space.

$$(6.7) \qquad \text{Đ}_{A,B} = C_{A,B} \cdot C_{B,A}$$

$$(6.8) \qquad \text{Đ}_{A,B} = \text{Æ}_{A,B} \cdot S_{rA}^{\alpha\beta}\left(Sit\,\eta_A, Sit\,\eta_B\right) \cdot \text{Æ}_{B,A} \cdot S_{rB}^{\beta\alpha}\left(Sit\,\eta_B, Sit\,\eta_A\right)$$

Given (6.9), we can also write (6.10).

$$(6.9) \qquad \text{Æ}_{A,B} = \frac{\left|E_A \cap E_B\right|}{\left|E_A\right| + \left|E_B\right|} = \frac{\left|E_B \cap E_A\right|}{\left|E_B\right| + \left|E_A\right|} = \text{Æ}_{B,A}$$

$$(6.10) \qquad \text{Đ}_{A,B} = \text{Æ}_{A,B}^2 \cdot S_{rA}^{\alpha\beta}\left(Sit\,\eta_A, Sit\,\eta_B\right) \cdot S_{rB}^{\beta\alpha}\left(Sit\,\eta_B, Sit\,\eta_A\right)$$

To shorten the equation, we introduce the *interdependence factor W* as the product of the two double sums of relevance factors, as shown in (6.11). Now, we can formulate the copula potential between the two finalities as (6.12) shows. The total copula potential can be formulated simply the sum of all copula potentials between the perspector and all objects we intend analyze their detectability.

In fact, we may formulate MS itself (or any other language) as a finality and everything else (or, if necessary for description, the language itself as well) as a sum of finalities with their attributes. Subsequently, oversight can be formulated as the copula potential with respect to the situative elements, whereas insight can described by the total copula potential considering solely situant elements ((6.13) and (6.14), respectively). The sum of both oversight and insight is equal to the whole copula potential of the given language, as shown in (6.15).

$$(6.11) \qquad W_{rA,rB}^{\alpha\beta}\left(Sit\,\eta_A, Sit\,\eta_B\right) = S_{rA}^{\alpha\beta}\left(Sit\,\eta_A, Sit\,\eta_B\right) \cdot S_{rB}^{\beta\alpha}\left(Sit\,\eta_B, Sit\,\eta_A\right)$$

$$(6.12) \qquad \text{Đ}_{A,B} = \text{Æ}_{A,B}^2 \cdot W_{rA,rB}^{\alpha\beta}\left(Sit\,\eta_A, Sit\,\eta_B\right)$$

$$(6.13) \qquad O(\text{Đ}_A) = \sum_{o=0} \text{Æ}_{A,o}^2 \cdot W_{rA,ro}^{\alpha,o}\left(Sit\,\eta_A, Sit\,\eta_o\left(Fin_{3,o}\right)\right)$$

35

$$(6.14) \qquad I(\mathrm{Đ}_A) = \sum_{i=0} Æ_{A,i}{}^2 \cdot W^{\alpha,i}_{rA,ri}\left(Sit\,\eta_A,\, Sit\,\eta_i(Fin_{\text{ч},i})\right)$$

$$(6.15) \qquad O(\mathrm{Đ}_A) + I(\mathrm{Đ}_A) = \mathrm{Đ}_A$$

## 6.4 Range

The range is an important property of affects, as it represents how 'far' the influence of a finality can reach. However, the range can trespass the edges of the situative space. In general, any space can be part of the range. Affects with an inter-space range are called *situaric affects*. We may formulate the *range R* as the sum of all situar elements which are in a not-nonvial relation with the regarded affect:

$$(6.16) \qquad R\left(Sit\varepsilon_A\right) = \sum_{v=0}^{w} E_{\text{Ц}} \,\bigg|\, Kon\rho := \left\langle Sit\varepsilon_A; E_{\text{Ц}}\right\rangle$$

If we relate this value to the number of situars which are known to the observer, we obtain the *relative range $R_{rel}(Sit\varepsilon)$*:

$$(6.17) \qquad R_{rel}\left(Sit\varepsilon_A\right) = \frac{\displaystyle\sum_{v=0}^{w} E_{\text{Ц}} \,\bigg|\, Kon\rho := \left\langle Sit\varepsilon_A; E_{\text{Ц}}\right\rangle}{\displaystyle\sum_{x=0}^{y} E_{\text{Ц}} \,\bigg|\, E_{\text{Ц}} \in \mathcal{K}_{obs}}$$

## 6.5 Affection

Since affects are implied by attributes, we can deduce the different types of affection from attribute combinations, because all affection processes can be formulated in the situant space (5.2). Accordingly, we will use the same notifications for the amounts of initial, added and annihilated attributes. As we intend to explain the combination of attributes as precisely as possible, we need to formulate the mechanism of affection in a more distinct way.

### 6.5.1 Relevance Vector and Relevance Factor

As has been shown in 6.2, the compatibility is a quantitative description of the influence of one finality can have on another. Still, we have not explained the relevance factor yet. In order to provide a satisfying explanation, we need to choose a subdivision of affection which the relevance factor can be derived from. If we recite the different combinatorial types, we can classify all of them as a combination of two affections: Either parts of attributes are annihilated (destruction) or new ones are constructed (construction). Hence, we can illustrate relevance as a vector in a two-dimensional Cartesian coordinate system. Let the amount of destructed attribtues be δ, and the amount of constructed attributes κ, as shown in (6.18). The destruction coordinate is negative, since it represents an attribute loss. Its minimum value is -*m*, else it is written as –*k* or zero, respectively. The construction coordinate is always positive, and its amount is, generally, arbitrary, which is represented by the variable *o*. Summing up, we can assign relevance vectors with specific coordinate values to each affection type (**chart 6.1**). The relevance factor can be interpreted as the value of the relevance vector and can be written as (6.19). If we are interested in the changing rate of attribute quantities, we can normalize the factor to the amount of initial properties (6.20). The relevance vector defines the combinatorial type of the according attribute. Thus, the values for destruction and creation coordinates are taken from **chart 5.1**.

(6.18)
$$r = \begin{pmatrix} \delta \\ \kappa \end{pmatrix}$$

(6.19)
$$\|r\| = \sqrt{\delta^2 + \kappa^2}$$

(6.20)
$$\|r\|_{norm} = \frac{\sqrt{\delta^2 + \kappa^2}}{\left| Sit\,\eta_{B1} \right|}$$

## 6.5.2 Affection Function

The affection can be interpreted as the outcome of a function, which can be derived from the situative matrices $Sit\sigma_1$ (6.21) and its successor $Sit\sigma_2$ (6.22). In both matrices, finalities are left out for simplicity. The object attributes are separated into two classes, the initial attributes $Sit\eta_{B1O}$ and the new, constructed attributes $Sit\eta_{B1N}$. Accordingly, $Sit\sigma_1$ has no constructed attributes. We can assign coordinate values to each type of affection and calculate the values for the successor situation (**chart 6.2**). Given this oversight, we can formulate the *affection fuction* $\Phi(Sit\varepsilon_{A,B})$, shown by (6.23).

$$
(6.21) \quad Sit\sigma_1 := \left\{ \begin{array}{l} Sit\varepsilon_A : \left( \begin{array}{c} Sit\varepsilon_{A1} \\ ... \\ Sit\varepsilon_{An} \end{array} \right) \\[2em] Sit\eta_A : \left( \begin{array}{c} Sit\eta_{A1} \\ ... \\ Sit\eta_{An} \end{array} \right) \quad Sit\eta_{B1} : \left( \begin{array}{l} Sim_{B1O} : \left( \begin{array}{c} Sim_{B1O1} \\ ... \\ Sim_{B1Om} \end{array} \right) \\ Sim_{B1N} : (\varnothing) \end{array} \right) \end{array} \right| Kon\rho_{A,B} \right\}
$$

$$
(6.22) \quad Sit\sigma_2 := \left\{ \begin{array}{ll} Sit\varepsilon_A : \left( \begin{array}{c} Sit\varepsilon_{A1} \\ ... \\ Sit\varepsilon_{An} \end{array} \right) & Sit\varepsilon_A : \left( \begin{array}{c} Sit\varepsilon_{A1} \\ ... \\ Sit\varepsilon_{An} \end{array} \right) \\[3em] Sit\eta_A : \left( \begin{array}{c} Sit\eta_{A1} \\ ... \\ Sit\eta_{An} \end{array} \right) \quad Sit\eta_{B2} : & \left( \begin{array}{l} Sim_{B2O} : \left( \begin{array}{c} Sim_{B2O1} \\ ... \\ Sim_{B2Op} \end{array} \right) \\ Sim_{B2N} : \left( \begin{array}{c} Sim_{B2N1} \\ ... \\ Sim_{B2Nq} \end{array} \right) \end{array} \right) \end{array} \right\}
$$

$$
(6.23) \quad \Phi\left(Sit\varepsilon_{A,B}\right) = \left| Sit\eta_{B1} \right| + \delta + \kappa
$$

It is important to note, that the relevance vector coordinates solely represent one situant effect of $Sit\varepsilon_{A1}$. If we apply the affection function on all situant affects, we can simply sum over all functions. However, situant attributes cannot be annihilated twice, thus the destruction coordinates must be concerted, represented by the index $C$, as shown in the final formulation for affection, (6.24).

$$
(6.24) \quad \Phi\left(Sit\varepsilon_{A,B}\right) = \sum_{i=1}^{x} \Phi\left(Sit\varepsilon_{A,Bi}\right) = \sum_{i=1}^{x} \left(\delta_C + \kappa_C\right)_i
$$

**Chart 6.1**: Affection types of affects and the according relevance vectors with their coordinates.

| Affection Type | Relevance Vector | Coordinate Values |
|---|---|---|
| tan | $\begin{pmatrix} 0 \\ \kappa \end{pmatrix}$ | $\delta=0$, $\kappa=o$ |
| pan | $\begin{pmatrix} -k \\ \kappa \end{pmatrix}$ | $\delta=-k$, $\kappa=o$ |
| trn | $\begin{pmatrix} -m \\ \kappa \end{pmatrix}$ | $\delta=-m$, $\kappa=o$ |
| tn | $\begin{pmatrix} -m \\ 0 \end{pmatrix}$ | $\delta=-m$, $\kappa=0$ |
| pn | $\begin{pmatrix} -k \\ 0 \end{pmatrix}$ | $\delta=-k$, $\kappa=0$ |
| n | $\begin{pmatrix} 0 \\ 0 \end{pmatrix}$ | $\delta=0$, $\kappa=0$ |

**Chart 6.2:** Affection types of affects and the according values for the successor situation $Sit\sigma_2$ according to (6.17). $|Sit\eta_{B2}|$ means the amount of elements of situant space of grade two.

| Affection Type | $\delta+m$ | $\kappa$ | $|Sit\eta_{B2}|$ |
|---|---|---|---|
| tan | $m$ | $o$ | $m+o$ |
| pan | $k$ | $o$ | $k+o$ |
| trn | $0$ | $o$ | $o$ |
| tn | $0$ | $0$ | $0$ |
| pn | $k$ | $0$ | $k$ |
| n | $m$ | $0$ | $m$ |

## 6.6 Activation

### 6.6.1 Different Types of Activation

Beside range and affection, it is important to classify the *activation* possibilities as well. Generally, if an affect occurs, we can differentiate three different cases: 1) The affect changes the object, as has been depicted in 6.5, whereas the affect is not transmitted. 2) The affect activates the object to induce a different affect by itself. 3) The affect activates the object to induce an affect back on the subject. In terms of MS, we can interpret the activation of an affect as an operator, which connects the initial affect with the activated one. Concluding from 1), an affect can be *deactivated*. This means that the object finality will not proceed to affect other finalities depending on the observed affect. In the second case, the object finality is affected to induce an affect (of same or different type) on other finalities. This process is called *transactivation*. As a special case of transactivation, we can define *reactivation* as transactivation towards the subject finality. In terms of scientific analysis, reactivation is the initial affection process leading to interdependence.

### 6.6.2 Activation Function

Similar to 6.5.2, we first need to quantize and structuralize the activation process. By formulating the function as a logical operator as has been said before, we can distinguish the types of activation by the scheme shown in **fig. 6.2**. The activation types as operators can be formulated by (6.25) for deactivation, by (6.26) for transactivation and by (6.27) for reactivation.

$$(6.25) \qquad F\left(Sit\varepsilon_{AB}\right) = \{\varnothing\}$$

$$(6.26) \qquad F\left(Sit\varepsilon_{AB}\right) = Sit\varepsilon_{BC}$$

$$(6.27) \qquad F\left(Sit\varepsilon_{AB}\right) = Sit\varepsilon_{BA}$$

Now that the frame conditions are known, we can quantify them by introducing the activation function $T$ which provides the according values:

$$(6.28) \qquad T\left(Sit\varepsilon_{AB}\right) = 0$$

for deactivation,

$$(6.29) \qquad T\left( Sit\varepsilon_{AB} \right) = \mathcal{Æ}_{A,B} \cdot S_{rA}^{\alpha\beta}\left( Sit\,\eta_A, Sit\,\eta_B \right) \cdot \mathcal{Æ}_{B,C}\, S_{rB}^{\beta\gamma}\left( Sit\,\eta_B, Sit\,\eta_C \right)$$

for transactivation,

$$(6.30) \qquad T\left( Sit\varepsilon_{AB} \right) = \mathcal{Æ}_{A,B} \cdot S_{rA}^{\alpha\beta}\left( Sit\,\eta_A, Sit\,\eta_B \right) \cdot \mathcal{Æ}_{B,A}\, S_{rB}^{\beta\alpha}\left( Sit\,\eta_B, Sit\,\eta_A \right)$$

for reactivation.

Generally, (6.29) must be rewritten to consider any possible finality, as the activated affect can, in general, aim more than one further finality. The according scheme is shown in **fig. 6.3**. Moreover, we can rewrite the equations by applying (6.6). This leads to equations (6.31) to (6.33).

$$(6.31) \qquad T\left( Sit\varepsilon_{AB} \right) = 0$$

for deactivation,

$$(6.32) \qquad T\left( Sit\varepsilon_{AB} \right) = C_{A,B} \cdot \left( C_{B,C} + C_{B,D} + C_{B,E} + \dots C_{B,n} \right) = C_{A,B} \cdot \left( C_{B,C} + \sum_{k=C}^{n} C_{B,k} \right) \qquad n \in \mathbb{N}$$

for transactivation,

$$(6.33) \qquad T\left( Sit\varepsilon_{AB} \right) = C_{A,B} \cdot C_{B,A}$$

for reactivation.

Indeed, it is possible that transactivation and reactivation occur at the same time. Summing up the equations (6.31) to (6.33), we can formulate the activation function:

$$(6.34) \qquad T\left( Sit\varepsilon_{AB} \right) = C_{A,B} \cdot \left( C_{B,A} + \sum_{k=C}^{n} C_{B,k} \right)$$

## 6.7 Combined Classification of Affects

We have developed three ways to classify an affect: By its range ($R$), by its affection ($\Phi$) and by its activation ($T$). Each property can be interpreted as one dimension of classification, thus we can sum them up into a three-dimensional function **K**:

$$(6.35) \qquad K\left(Sit\varepsilon_A\right) = \begin{pmatrix} T \\ \Phi \\ R \end{pmatrix}$$

Finally, we can easily formulate the explicit form by applying the formulations (6.16), (6.24) and (6.34), obtaining a three-dimensional classification of affects.

$$(6.36) \qquad K\left(Sit\varepsilon_A\right) = \begin{pmatrix} \sum\limits_{x=B}^{y} C_{Ax} \cdot \left( C_{xA} + \sum\limits_{k=x}^{n} C_{x,k} \right) \\[2ex] \sum\limits_{x=B}^{y} \sum\limits_{i=0}^{x} \left| Sit\,\eta_{xl} \right| + \delta_{C,i} + \kappa_{C,i} \\[2ex] \sum\limits_{v=0}^{w} E_{II} \,\Big|\, Konp := \left\langle Sit\varepsilon_A; E_{II} \right\rangle \end{pmatrix}$$

**Fig. 6.2:** Affections of finalities A, B, C, represented by arrows.

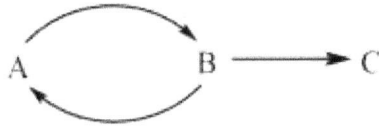

**Fig. 6.3:** Extended version of the scheme shown by **fig. 6.2** with $n$ objects of the activated affect.

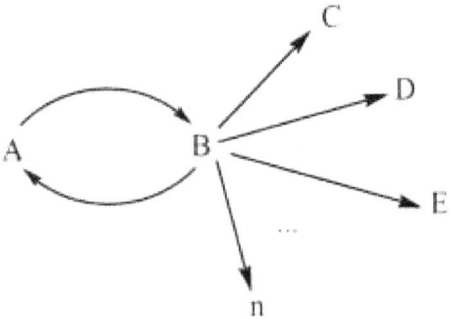

# PART 2:
## TRANSFORMATION OF MATRICAL SITUARICS INTO RECOMPLEX TENSOR SEQUENCE
## DEVELOPMENT OF A PARADIGMATIC ALGORITHM

## 7. Recomplex Tensor Sequence [12, 13]

After the development of MS syntax in PART 1, we are able to describe various situations depending on perspective, precision and point of interest. The introduced instruments can be used to elaborate new insight about research, which will show us both abilities and limitations of the MS as well as allow various applications. Mathematical and physical applications, for instance, will be discussed in later works, as some crucial aspects still are to be elaborated. In order to roughly depict the intended orientation, chapter 10 will briefly illustrate the according ideas. PART 2 will focus on an alternative language, the *recomplex tensor sequence*, hereinafter RTS. Furthermore, an application in science theory will be discussed, as has been mentioned in 1.3.

The RTS is a re-organized formulation of the matrical situarics. Its intention is to provide a language according to the postulates, which can be interpreted as or integrated into a programming language. Against the background of the development of artificial intelligence, an increasing, worldwide information flow and ever-increasing complexity of scientific models, it is crucial to have a common basis on which pragmatic and fast solutions can be elaborated in any scientific branch. Basically, it represents an alternative approach according to the postulates to fulfill the manifold requirements following the depicted circumstances mentioned above.

### 7.1 Recomplexation as Translation

According to the integrity principle, a situative matrix is defined by its structure and relation with respect to the whole situaric space. If any detail is changed, the integrity changes and thus the situation changes as well. This means, that a structure undergoes a rearrangement during an analysis process. As we intend to formulate a set of laws to create a new language, we need to rearrange the information given by the MS structures into a new structure. The translation is not a rearrangement within a given syntax, but a rearrangement to *obtain a new syntax*. In order to differentiate between both types of rearrangements, the translation process is called *syntactic recomplexation* or just *recomplexation*.

Since it is a crucial advantage to have a recapitulatory sight on the development of descriptions, we sum up the ever-changing situations within *tensors*. From our point of view, this is a recomplexation. Thus, it will is called a *recomplex tensor*. To enable the syntax to create sub-groups, parallel processes and meta-analysis, recomplex tensors can be formulated as parts of larger structures, which are called *recomplex tensor sequences*, or just *tensor sequences*. With an ever-increasing interlation process, and according to the postulates, everything can be summed up into one single tensor sequence, therefore the language's name.

Since the postulates require both languages to be able to describe each other, the MS must be able to describe both the recomplex tensor and the recomplex tensor sequence itself as a situation. Indeed, this can be achieved by formulating each element of the syntax as a finality with its attributes and affects and connecting the elements accordingly with constructed situations. To sum up the situations, we introduce new constructed situations. A recomplex tensor is called *Kon $\tau$*, a recomplex tensor sequence is written as *Kon $\theta$*.

## 7.2 Hierarchy Function

Now that the general relations have been clarified, we need to formulate the detailed laws for the syntax of the RTS. Since we start off from MS, we will keep the elements and constructed situations as elements of a syntax and simply rearrange them.

In order to do so, we introduce the *hierarchy function H*. This function assigns a defined position within a hierarchy for any element, constructed situation and even spaces, subsequently creating a new arrangement. As spaces define the structure of the syntax, they need to have the highest rank within the hierarchy. The situaric space occupies the highest rank among them, as it is the all-engulfing space. It is followed by the situatoric, the situative and, finally, the situant space. As spaces can be interlaced within each other, the rank of an interlaced space is always lower. These requirements are represented by the hierarchy function laws 1) and 2), which are formulated in 7.3.

Furthermore, we need to clarify the hierarchy between the different elements and constructed situations. In general, constructed situations applied on elements of higher rank also occupy the higher rank. Hence, the application of functions on both tensoric sequences and recomplex tensors must have a higher rank than the constructed situations themselves, which results in functions of tensoric sequences occupying the highest rank. Moreover, all constructed situations introduced in chapter four have a lower rank than the basal elements, since they provide the basal information and thus need to be formulated first. The highest rank among the remaining constructed situations is occupied by functions, as they can be applied on all other constructed situations. They are followed by operators, perspectives and finally, relations. Furthermore, constructed situations of the same type are ordered according to the rank of their arguments. These requirements are formulated in laws 3) and 4).

If more than one perspective is enlisted within a structure given by law 3), the differentiation occurs with respect to the absolute value of their copulae's elements, as shown in law 5). Law 6) clarifies the hierarchy between affects, attributes and finalities.

## 7.3 Hierarchy Function Laws for Recomplexation

Summarizing the requirements for the hierarchy function depicted in 7.2, the six laws can be formulated as follows. They are to be applied in the following order:

1)    $H(Ц) > H(С) > H(З) > H(Ч)$

2)    $H(X^n) > H(X^{n-1})$

3)    $H(Kon\phi(Kon\theta)) > H(Kon\theta) > H(Kon\phi(Kon\tau)) > H(Kon\tau) > \dots$
      $\dots > H(E_С) > H(E_З) > H(E_Ч) > H(Kon\phi) > H(Kon\omega) > H(Kon\pi) > H(Kon\rho)$

4)    $H(Konx(X^n)) > H(Konx(X^{n-1}))$

5)    $H(Kon\pi_1) > H(Kon\pi_2) \,/\, |\mathcal{K}_1| > |\mathcal{K}_2|$

6)    $H(Sit\varepsilon) > H(Sit\eta) > H(Fin)$

## 7.4 Orientation and Formulation

After the recomplexation laws are formulated, we can proceed to formulate the actual syntax laws of RTS.
In general, the language shall be orientated like a programming language. This means, the syntactical orientation is mainly vertical, whereas the horizontal position has a secondary importance. We differentiate between the structure of a situative matrix and other notifications, in which constructed situations are formulated. The Greek letters which symbolize the different types are replaced by declarations written in Latin letters, as shown in **chart 7.1**.

### 7.4.1 Space Shift

The horizontal orientation is provided by the spaces, which are aligned according to their hierarchy from left to right and are separated by a shift. Thus, there are space columns in which the notifications of according elements of the syntax begin. Situars begin in the second column, situators in the third and so on.

As long as they are not elements of described situative matrices, tensoric sequences are written in the first column. Operators and functions which applied on tensors, sequences or situators are written in the second space of the situaric column, as well as tensors which are not elements of known situative matrices.

Situators, which describe a situative matrix, are written in the situatoric column. All situators describing a matrix in particular are followed by the announcement `structure` in the second space of the situator column, after which the elements of the matrix are enlisted. After the structure has been described, the announcement `structure end` is written and followed by constructed situations in the second space of the situator column, as far as they are known. If the matrix contains situators, these are enlisted in the situative column, as they are sub-situators and are ordered in the lower rank. Their structure and constructed situations are all shifted accordingly. In general, if syntax elements have any description or sub-elements, they are first announced, followed by the description shifted by one space or sub-elements shifted by one column and end with their notification, added by an `end`, subsequently. This notation is used for any constructed situation as well.

During the `structure` sequence, elements enlisted within a line are separated using semicolons. Outside the `structure` sequence, they are separated by commas.

## 7.4.2 Functions and Operators

For both functions and operators, the first line after their declaration is the argument line, whereas the text is shifted by one space with respect to the declaration. It begins with the announcement `of`, followed by the argument. If there are many arguments, they are enlisted and separated by commas, as has been stated in 7.4.1. If there are no known arguments or the amount of arguments is open, the `of` line can be left out.

If-structures, repetitions and loops can be formulated as operation orders. Since RTS is yet to be developed into an actual programming language, one can include every possible program language syntax. However, setting of syntax rules and development of an according compiler is the subject of further research.

## 7.4.3 Perspectives

Perspectives are anounced and ended like functions and operators, whereas the `of` announcement is required to clarify which finality's perspective is meant. In the second line, all elements of the copula are enlisted. All lines in between the announcement of the perspective and its ending are shifted by one space. Element announcements are formulated as shown in 7.4.5.

### 7.4.4 Relations

Relations have a line for their subjects, followed by a line each for every object type two, and a last one for type one objects. They are shifted by one space with respect to their annoucenement, similar to the other constructed situations.

### 7.4.5 Element Annoucnements

Whenever we do not need to formulate a structure of a situative matrix or, simply, want to formulate shortly that we include all elements of a given space, we can shortly announce all elements of the according space, writing E[SPACE,e[SPACE](META-ELEMENT)]. The minor letter e represents the variable which enlists every single element within the according space. Space parts are treated like simple quantities, thus as arrays.

### 7.4.6 Enlisting Variables

As mentioned in 7.4.5, we can enlist every element of a quantity by a variable. If we have, for instance, $n$ sub-finalities of the meta-finality $x$, we can vary the variable $m$ from one to $n$. The notification is Fin[x,m], whereas the variable $m$ has to be defined before: m[1:n]. This notification is chosen to shorten the syntax.

### 7.5 Application

In order to illustrate the shown hierarchy function laws and syntax rules, we need a simple example of application. The situative matrix shown in (7.1) may be used for this purpose. Single minor space relations are interpreted as basal relations between finalities and their attirbutes, and the situaric relations as relations between finalities of the different situators. Only one of them will be formulated. The operator connects $Finx_1$ and $Finx_2$, whereas $Kon\phi$ is the $\mu$-function as described in 2.3.3. The shown perspective may belong to $Finx_1$, which copula contains all elements within $Sit\sigma_R$.

$$(7.1) \qquad Sit\sigma_R := \left\{ \begin{array}{l} Sit\varepsilon_1 \\ \\ Sit\eta_1 \quad Sit\eta_2 : \left( \begin{array}{l} Sit\eta_{21} : \left( \begin{array}{l} Sit\eta_{211} \\ Sit\eta_{212} \end{array} \right) \\ \\ Sit\eta_{22} \end{array} \right) \\ \\ Finx_1 \quad Finx_2 : \left( \begin{array}{l} Finx_{21} : \left( \begin{array}{l} Finx_{211} \\ Finx_{212} \end{array} \right) \\ \\ Finx_{22} \end{array} \right) \end{array} \right| \left. \begin{array}{lll} Kon\tau & Kon\pi & Kon\rho_y \\ Kon\omega & Kon\rho_{\amalg} & Kon\rho_{y2} \\ Kon\phi & Sit\sigma_S & Kon\rho_{\amalg'} \end{array} \right\}$$

47

with

$$Kon\rho_{II'} = \mu(Kon\rho_{II})$$

and

$$Sit\sigma_S := \left\{ \begin{matrix} Sit\varepsilon_3 \\ Sit\eta_3 \\ Finx_3 \end{matrix} \right\}$$

The according recomplex tensor sequence is formulated as follows.

**Chart 7.1:** Notifications for basal situar elements and constructed situations in MS and RTS.

| Notification in MS | Notification in $Kon\theta$ |
| :---: | :---: |
| $Sit\sigma_x$ | Sits |
| $Sit\varepsilon_x$ | Site |
| $Sit\eta_x$ | Sith |
| $Finx$ | Fin[x] |
| $Kon\theta_x$ | Konq[x] |
| $Kon\tau_x$ | Kont[x] |
| $Kon\phi_x$ | Konf[x] |
| $Kon\omega_x$ | Know[x] |
| $Kon\pi_x$ | Konp[x] |
| $Kon\rho_x$ | Konr[x] |

```
>Kont[7.1]
     >SitZs[R]
      >structure
          >Sits[S]
           >structure
                >Site[3]; Sith[3]; Fin[x,3]
           >structure end
          >Sits[S] end
          >Site[1]; Sith[1]; Fin[x,1]
          >Sith[2]; Fin[x,1]
                >Sith[21]; Fin[x,21]
                     >Sith[211]; Fin[x,211]
                     >Sith[212]; Fin[x,212]
                >Sith[21]; Fin[x,21] end
          >Sith[2]; Fin[x,2] end
      >structure end
      >Kont
      >Konf
       >of Konr[Ц]
       >Konr[Ц']
      >Konf end
      >Know
       >of Site[1], Sith[1], Fin[x,1]
       >Site[2], Sith[2], Fin[x,2]
      >Know end
      >Konp
       >of Site[1], Sith[1], Fin[x,1]
       > E[Ц,e[Ц](R)]
      >Konp end
      >Konr[Ц]
       >Site[2], Sith[2], Fin[x,2]
       >Sith[3], Fin[x,3]
      >Konr[Ц]end
      >Konr[Ц']
      >Konr[Ч]
      >Konr[Ч^2]
     >Sits[R]end
>Kont[7.1]end
```

# 8. A New Concept in Sicence Philosophy [7, 8, 10, 11, 21 - 27]

## 8.1 Structure and Dynamics of Scientific Theories

After RTS has been developed as an algorithm-adaptive version of the matrical situarics, we can apply this extraabstract language on different scientific branches. Instead of applying RTS on each scientific branch, we can simply apply it on science philosophy, hereinafter SP, as it includes all branches. If we are able to formulate a recomplex tensor sequence which describes SP, we can, subsequently, use the general scientific pattern and elaborate a description for each branch. First, we need to solve several problems of the current conception of SP.

Before we elaborate a new concept of scientific development, it is important to recapitulate the historical development of SP itself. This work focused on the concepts depicted by Ulrich Gähde in 'Science History' by Andreas Bartels and Manfred Stöckler. Gähde uses three examples to describe the development of ST: Karl Popper's fallibilism, Imre Lakatos' methodology of research programmes as well as the structuralistic approach, which has been developed by Sneed and other authors. A brief characterization of the structuralistic approach will be used as a starting point. This approach has been chosen because it has been tested and improved numerous times on concrete scientific theories. Hence, it provides the best expectations for descriptions with high precision.

The structuralistic concept orders theories in a specializing structure, in which the basic theory, $T_0$, or the basal axiomatic structure is added by specialized theory elements, $T_i$, which are applied for special situations. $T_0$ is called the theoretical *core*, whereas all $T_i$ are elements of the so-called *protective belt*. Each theory $T$ is a well-defined as a pair of a *model M* and an *application quantity I*. The application quantity of $T_0$ is decreased by its specializations, which form a net of interlacing specializations and accordingly specializing models. This forms the tree diagram-like structure, which endings are the most specialized additional theories. Since theories depend on their application quantities, the structure can be deduced from data structures which specialize databases. These provide, in turn, the definition basis for application quantities. Whenever there is a phenomenon which cannot be described by the current structure, the protective belt is always modified before the core. The modification must occur minimally invasive. This means, as many as possible elements of the structure must remain unchanged after the modification. Thus, the most specialized theories are focused first.

There are three possible modifications. The first is an addition of even more specialized theories. This implies a growth of the protective belt. The second possible modification is replacement, where a single specialized theory is replaced by another. As a third possibility, specialized theories can be simply erased. Only if the belt is erased completely by modifications, the core undergoes the modification process. As long as possible, it must remain unchanged.

50

In general, models have the role of mediators between reality and theory, but the actual relation between models and data structures as well as theories remains unclear in structuralism. Furthermore, the formation of both models and theories has not been clarified yet.

## 8.2 Sequential Approach for Science Philosophy

The requirements for a new concept include a more precise description of model and theory formation, as has been said above. Furthermore, the new concept is intended to be a development of the steadily applied and improved structuralistic concept. As the formulation of the new concept will aim to fit into a tensoric sequence, it is to be referred to as *sequential approach* or *sequential concept* of SP. First, the mentioned problems will be solved. Afterwards, several other aspects will be discussed and adapted to the whole concept. Subsequently, the macrostructure of the concept will be developed into a far more general version, which will include the aspects mentioned in 1.1.1 and verify the satisfaction of requirements formulated in 1.1.2.

### 8.2.1 Development of Models and Application Quantities

Both models and application quantities are formulations connecting theoretical statements with reality. In order to derive the relations between them, we need to clarify how models and application quantities are created in the first place.

Science grounds on observation and so does the development of theories. Hence, models have to emerge from a process which starts with observation. The very basic observation of an event is noticing the properties and behavior of objects within a confined range of space and time. Without any theory in the beginning, this is the very first step of science.

Since human-like observers eager to explain which types of events occur under given circumstances, we organize the objects by their mutual properties. This organization presumes the understanding of the concept of similarity. This is why we elaborated the basal similarity grade in 5.1. In general, we can focus solely on attributes, as they store information which is the main subject of model development. The unspecified finalities which obtain their character by attributes are, in this context, manifestations of the same type of observed behavior.

In the next step, all mutual properties are classified in order to formulate a conclusion. The classification depends on the current theory during model development. If there is no theory yet, the classification process can be arbitrary. In further development, the classification will either lead to successful formation of theories or fail. In case of failure, the classification was wrong and the process must start from beginning. If there is a theory already, the classification depends on the theory's application quantity and must imply a subset of it. Otherwise, the theory structure described in 8.1 could not be achieved. It is important to note that by following the application quantity, one must also apply the theory's jargon, which is entangled with the theoretical interpretation or, in our terms, classification.

After the classification, the next step may be described as *guessing*. As simple this term seems to be, it is the crucial step in theory development. Instead of 'guess', one can also use the term 'hypothesis'. 'Guessing' means to formulate a causal relation between two groups of classes, a subject group and an object group. Given the existence of objects of type two introduced in 2.3.3, one could say this concept may be too simple. In fact, the differentiation between objects of type one and type two is solely the development of a model into a more sophisticated one, in which objects type two are subjects of a relation which still aims to objects of type one. Thus, the classification into subjects and objects describes the guessing step with sufficient precision despite its simplicity.

In the next step, the guess has to be verified. This can be regarded as an *experiment type one*. The experiment focuses on the observation from the beginning. If the guessed causal relation explains the observation properly, the guess was right and can be evolved into a theory. If not, another guess must be made. The condition for the new guess must be that it is different from all previous ones. If there is not possibility left to formulate a new, different guess, we may either assume that the initial conclusion was wrong or the amount of information in the observation was not sufficient to formulate causal relations, which means that a new observation must be done. To err on the side of caution, the last option shall be taken. In that case, the whole process starts from beginning – with a new observation providing more information.

If the guess was right, the theory can be developed. First, its elements need to be obtained. At this point it is important to note that the application quantity can be derived from the model itself. Let us define a model as a system in which the observed objects have defined properties under defined circumstances and undergo defined modifications. Therefore, the application quantity is defined by the properties, circumstances and modifications mentioned in the according model. We can sum up all three as the properties of a given system. According to the postulates, a system can be defined as a situation. After applying the Möbius curvature, a situation can be formulated as a finality with defined attributes. These attributes are the properties of the system described before. As the model grounds on the successful guess, its objects must be of the same type as the classification of the guess. Finally, we can conclude that the subjects mentioned by the guess define the application quantity, as the model can only be applied on subjects described by the formulated attributes. Hence, both the model and its application quantity can be developed simultaneously from the successfully guessed causal relation between subjects and objects.

In order to develop the application quantity, we simply formulate the condition that all subjects of the model must have properties under which the model expects the effect mentioned in the guess. The model itself includes not only the subject condition but also the object condition, which describes the properties of the objects. Furthermore, it describes the causal relation between subject and object, which viality aspects can be regarded as the according laws of nature.

## 8.2.2 Formulation of Scientific Theories

Similarly to the structuralistic approach, we interpret theories as a defined pair of model and application quantity. Thus, we can formulate a relation between both elements which defines the according theory. If we formulate a situation in terms of MS which includes elements and the relation between them, this situation will represent the scientific theory. As we need to treat it as a theoretical object, we apply the Möbius curvature and obtain a situative element, whereas both model and application quantity are its sub-elements. The relation between them simply mutates into a situant one.

In general, we can interpret models as finalities and application quantities as according attributes. Indeed, other properties of models shall be formulated as attributes as well. Since we focus on application, we can cancel out other attributes from our copula. Furthermore, this approximation prevents redundant meta information.

## 8.2.3 Verification of Prediction

Given the formulation of 8.2.1, prediction of scientific theories is already provided by the structure of models. If subjects of the described type are found to have attributes required by application quantity and occur under circumstances formulated by the model, they behave according to the causal relation. However, experiments for verification of scientific prediction must be chosen according to criteria provided by the application quantity. All events which elements satisfy the requirement given by the application quantity can be used as experiments. This loose condition allows further development of theories, as slight divergence implies additional information, which increases the range of the scientific theory by forcing it to spread. In contrast, strict conditions would simply specialize the whole theory into a small range of natural events, which would restrict the theory's perspective.

If the experiment follows theory's prediction, it can be included into the amount of events which have been described by the theory already and the next experiment can be chosen. If not, the theory has failed to describe the outcome of the experiment and the protection belt has to be elaborated, which is the subject of 8.2.4. Generally, experiments aiming to verify the prediction of an already constructed theory can be regarded as *type two experiments*.

## 8.2.4 First Element of the Protection Belt

Similarly to the first theory element or the basal theory, every new element of the theory structure is evolved the same way as depicted in 8.2.1. The event from which the first element of the protective belt is to be evolved is provided by the experiment the basal theory failed to explain. As mentioned before, the classification which leads to the conclusion is defined by the present basal theory. The classification properties must lie within the application quantity of $T_0$, otherwise the protection belt would not be built around the basal theory it has to protect. A successful guess, in this process, is evolved into a theory separated from the basal one. The belt is formulated as $T_i$, whereas the basal theory is part of $T_0$. Hence, $T_0$ and $T_i$ are situatives which sub-elements are the actual theory elements of the theory structure. The structural belonging of the single elements is defined by the according application quantity.

If the formation of a belt element is not possible, we may assume the same reason as depicted in 8.2.1. Accordingly, we need more information and thus a new experiment. It must describe an event which includes the current experiment in order to increase the required information. If the belt is formed, the next experiment can be chosen solely according to the application quantity.

## 8.2.5 Modification of Protection Belt and Core

Similar to Imre Lakatos' methodology of programmes, main process of theory development is described by the modification of its belt in order to protect the core. As have shown in 8.1, the structuralistic approach differs between three possible modifications: Erasement, addition and replacement. Each modification first aims the belt. In the new approach, all three modifications must occur simultaneously, whereas the best outcome is chosen to be the accepted modification. By 'best outcome', we may interpret the modification outcome which describes most of the experiment. The Darwinian approach is chosen in order to guarantee that any scientific development aims for improvement. If there are no belt theories left, addition is the only possible modification and a comparison is not required. How many times a modification loop can be repeated without success before the belt is abandoned, depends on many factors, as the effort and previous success of the modified theory structure. In general, one can set a maximum number for each loop. After both the belt and the core have been modified without success leaving not a single theory which can explain the experimental event, the whole process starts from beginning, in which the conclusion from the experiment is independent of any theoretical jargon, but represents a new perspective. In this case, the former theory failed and has to be replaced by a new one.

## 8.2.6 Science Branches

The described development of scientific theories depicts a single branch of science, such as physics, chemistry or biology. To enlarge the perspective, we can generalize this process on a given number of branches, which evolve out of different observed events. This means, that the scientific theories must be developed first, before they are organized into branches. We can interpret branches as single theories which undergo unification processes, occupying the application quantities of all single theories which have been fused. As has been stated in 1.1.1, the evolution of science needs to be convergent rather than divergent. Therefore, the compatibility with other theories events must be tested every time a theory is evolved, using any possible chance of decreasing the number of branches. If two theories are not compatible, their programs proceed to the next step, and the test repeats subsequently. If they are compatible, they merge into a unified theory. This process can be regarded as a *cross test*.

As an approach for a precise formulation of the cross test, we may first combine the application quantities of the two theories we intend to unify. In the next step, we prove whether simple combinations of the according application quantities can provide the combined application quantity. There are four possibilities: The first two cases occur if the combined application quantity is in fact one of the initial ones. In the third case, the intersection of both quantities is equal to the combination. In the fourth case, the new quantity cannot be described by the previous cases. This means, a combination of both quantities is the new application quantity. In either case, the models for the new theories are chosen accordingly. A fifth case occurs, if the combined quantities contradict each other. In this case, a new model must be found which provides the combined application quantity. At the first sight, this seems to be impossible. However, the contradiction is implied by the theories' understandings, which are provided by their models. Before we declare the cross test as failed, we must try to evolve a new model which, possibly, overcomes this contradiction or *incommensurability* (1.1.1). The according model must be adapted to the combined quantities. Indeed, this seems to be an information problem: Since the quantities contain less information about the theory than the model, as has been mentioned in 8.2.1, it is crucial to elaborate a process which elaborates the missing information. The required process may be called *reversed model research* or *reversed research of models*, hereinafter RRM. In fact, the RRM process is mostly similar to the usual development described in 8.2.1, with the crucial difference that the subject attributes are already known. Thus, the conclusion process must solely find according objects of the given event, which is represented by the fusion of both theories' successful experiments. In the guess-making, the according operators are yet to be found. Only if the whole process fails to find a common model, the cross test can be declared as failed on the given theory pair.

If the cross test fails on the whole theory set for a given number of times, new theories must be created from new observations. In the ideal case, in the end, there is only one theory left as a unification of the whole set of theories. It undergoes the iterative test with ever new experiments and, if necessary, the modification process depicted in 8.2.5. If this theory fails on an experiment, new observations of that type are required in order to create a new set of theories, which will evolve conclusions independent of the previous unified theory. The new set will include the unified theory as well as new theoretical approaches, which will then undergo the same process as the set before. In this way, the aim to formulate a unified theory is set as well as the acknowledgement, that new approaches are crucial for ever-continuous paradigmatic evolution.

### 8.2.7 Evolution Trends of Theory Sets

When the evolution of a set of scientific theories starts, there are three possible trends. A *divergent trend* occurs when the amount of theories increases due to the mechanism described in 8.2.6. Accordingly, a decreasing amount of scientific theories represents a *convergent trend*. Finally, a *stagnating trend* occurs when the number of theories does not change perceptibly over a period of loops. Additionally, the *evolution rate* can be calculated as the theory amount's first derivation with respect to time.

# 9. Development of PARTS [28, 29]

In this chapter, the sequential approach will be formulated as a recomplex tensor sequence in order to apply it as an algorithm. Generally, we can distinguish three levels of the whole concept. The *macrostructure* is the implemented general structure shown in 8.2.6. It decides which tensor sequences are applied under given circumstances. It is represented by the tensor sequence `Konq[Paradigmatic algorithm]`. The *mesostructure* arranges the structure of single sequences which can be interpreted as subsequences. The *microstructure* explains the formulation of single situators, operators and functions as well as the application of orders and special formulations, which will be briefly presented in 9.3.1.

### 9.1 Macrostructure

Sequences which represent the parts of the concept shown in 8.2.1 – 8.2.5 are summed up into the theory development sequence, `Konq[T,t]`. The variable t varies from one to $\theta$ and represents every single theory from a theory set. Furthermore, the cross test with RRM depicted in 8.2.6 also represents a sequence on its own, which is called `Konq[Crosstest(tau1,tau2)]` with $\tau_1$ and $\tau_2$ being two theories from the theory set. As shown in 8.2.6, the modification of theories is also applied on the last remaining theory from a theory set if it fails to explain an experiment. This means, a separated formulation of the modification sequence `Konq[Ex4]` has to be implemented. Additionally, the single modification steps (erasement, addition and replacement of theory elements) are formulated in a sequence, since they are applied differently with respect to the initial theory and modification development. Hence, they are summed up in the sequence `Konq[Rebuild]`. Finally, the calculation of a theory set's evolution depicted in 8.2.7 is implemented as an additional sequence `Konq[TrendCalc]`. While the theory set undergoes cross tests, the failed attempts in which no fusions can be undertaken are counted. After a maximum amount of failed fusions, new branches must be added to the set. Else, the sequence could run forever without actual improvement. **Fig 9.1** shows the macrostructure as a diagram (trend calculation is not included as it does not affect the theory set development).

Before the explicit formulation starts, another sequence must be formulated. It states all integers, real numbers, characters as well as variables. Furthermore, it formulates basal definitions of priority relations and the entanglement of finalities and their according attributes, which states that relations between finalities are projected automatically to attributes without explicit formulation.

The whole sequence is called *PARTS* (*P̲aradigmatic A̲lgorithm based on R̲ecomplex T̲ensor S̲equence*). Henceforth, this acronym will be used to describe the whole algorithm.

## 9.2 Mesostructure

### 9.2.1 Konq[T,t]

Theory development is the most important sequence of PARTS. As shown in 8.2, the whole process of a single theory consists of five steps:

1) Formulation of models and application quantities

2) Theory development

3) Application of basal theory

4) Development of the protective belt

5) Application of theory net (core and belt) and modification

The first step consists of conclusion and guess making, which are summed up in a tensor, Kont[1]. Afterwards, the elaborated guess is tested in an experiment of type one, which can be formulated in the tensoric function KonF[Ex1]. Henceforth, the functional formulation will be applied for all experiments. If the guess successfully describes the experimental result, both model and application quantity are developed by the tensoric function and the theory is built in another tensor, Kont[2]. The first experiment of type two, KonF[Ex2], proves the basal theory's ability to describe experiments depicted according to its application quantity. This experiment is iterative as long as the description is successful. If the basal theory fails on KonF[Ex2], it is to be modified by Kont[3] which develops the first element of the protective belt, following the same scheme as Kont[1]. Similarly, the belt guess is tested by an experiment of type one, KonF[Ex3]. As has been stated in 8.2.4, a failed belt guess making is followed by a direct application of the next experiment of type two, KonF[Ex4], in which a failed description is followed by modification process. Else, the successful guess is developed accordingly into a theory element of the protective belt by Kont[4], which subsequently leads to KonF[Ex4]. The mesostructure of Konq[T,t] is shown by **fig.9.2**.

### 9.2.2 Konq[Ex4]

Since the macrostructure requires a special treatment of the last remaining theory of a theory set, the fourth experiment of Konq[T,t] is formulated additionally as a sequence, whereas the microstructure is changed slightly. In general, the concept remains the same as depicted in 8.2.5.

### 9.2.3 Konq[Rebuild]

In order to formulate the modification process of a theory, Konq[Rebuild] is separated in six tensors, each three for the belt and core, representing the three possible modifications. Depending on the requirement given by the according experiment and the amount of belt or core theories, either solely the addition, replacement and addition or all three modifications can be applied. The erasement tensor is called Kont[NiBelt] or Kont[NiCore], respectively. It is the simplest modification, as it solely orders theory elements with respect to their priority and eliminates the one with the lowest priority. The declaration of priority in particular will be shown in 9.3.2.

The addition tensors, Kont[AdBelt] and Kont[AdCore], include the same scheme as Kont[1], as they develop a new theory element. Accordingly, they also include an experiment of type two which validates the guess. As the addition tries to compensate the insufficiency of the weakest element of the theory structure, it is combined with it in the end.

The replacement tensors, Kont[ReBelt] and Kont[ReCore], are combinations of erasement and addition tensors. First, the weakest theory is erased, thence a new one is added. This process does not combine the additional element with another one, but solely integrates it into the structure, whereas the application quantity is implied by the crucial experiment, which indicates the new theory element being an appropriate replacement. If more than one modification is done, the outcomes are ordered after their explanation power with subject to the crucial experiment. The outcome describing the experiment in the best way is chosen for further development.

**Fig. 9.3** shows the depicted mesostructure of Konq[Rebuild]. The mesostructures of different modification types are described in **fig. 9.4**. The explicit description is provided by 9.3.4.

### 9.2.4 Konq[Crosstest(tau1,tau2)]

The cross test of two different elements of a theory set undergoes the scheme described in 8.2.6, whereas the RRM process consists of an according formulation of Kont[1] and KonF[Ex1]. **Fig. 9.5** shows the structure of this sequence.

### 9.2.5 Konq[TrendCalc]

The calculation of theory set evolution trend of Konq[Paradigmatic algorithm] is done each time $\theta$ could change. The sequence includes the calculation of the relation between the old and the new amount of theories of the set as well the calculation of the change rate, as far as it is possible. Both values are written in a protocol to show the whole evolution. The explicit formulation is shown in 9.3.

## 9.3 Micro Structure

The microstructure of PARTS describes the formulations in particular. The described parts will be shown in the actual algorithm which is included in PART 3. Line numbers of the described parts are given in parentheses.

### 9.3.1 Special Expressions

As has been shown in 7.5, situative matrices are formulated as structures with their according constructed situations afterwards. Operators and functions can be formulated as if-structures and loops, as is first shown in Konq[Application] (2 – 185). The expression 'without' is simply declared by a backslash (298), followed by the element to be left out. 'Unequal to' is expressed by \= (299). The expressions and and or are operators used for quantities which are applied in Konq[Crosstest(tau1,tau2)], for instance (1500 – 1857). Furthermore, the operator norm can be used (1506), which means that no double occupations are allowed in the mentioned quantity. This application is simply used to prevent redundant data. The operator ALL is applied as well and means that all types of the according element are included (188, for instance). The function Class, or its plural equivalent Classes (319 for the first time) are classifications which depend on the according theory and its application quantity, as mentioned in 8.2.1. If there is no developed theory yet, the classification is random. Furthermore, the Möbius curvature operator is simply formulated with an overwriting announcement, showing which space the initial elements are to occupy. There are either positive Möbius curvatures for occupation of smaller minor spaces or negative curvatures for occupation of larger minor spaces (483 for instance). Descriptions are declared by description relations (163 – 166). Notation is done in the index and includes the declaration Descript, followed by the subject and the object indices. If the viality of a description relation is to be calculated, the `viality` function can be applied. For the amount of possible connections, the `abs` function is applied on the `viality` function, respectively (177, for instance). Further special formulations will be explained in discussion of according sequences. In general, each situation includes a perspective of the observer (for example 289 – 292) as well as a description relation towards the current relevance, if it is possible (413 – 416, for instance).

## 9.3.2 Konq[`Application`] (2 – 185)

Before the entanglement and priority relations are explained, all characters, real numbers, integers and variables are introduced (3 -126). Additionally, the input of the maximum iteration steps for modifications is required (34), as has been discussed in 8.2.5. Several variables are introduced later within loops in order to keep their declaration within the according sequence. After the sequence, KonF[`Application`] (186 – 189) declares that the depicted relations are to be applied on all sequences.

## 9.3.2.1 Kont[`Entanglement`] (127 – 146)

Subsequently, the entanglement tensor is formulated. The depicted situation includes an effect as well as a finality and its attribute (129 – 132). The tensor simply formulates that if an affect is aimed at the finality, it aims automatically at its attribute as well. Since affects are generally defined as the applications on elements in general, this applies on every possible change of any finality.

## 9.3.2.2 Kont[`Priority`] (147 – 184)

This tensor includes the declaration of counter variables as well as crucial relations between two arbitrary variable values (148 – 153). The following situative matrix includes a sub-situator which is described by models and their application quantities. In line 168, the priority counter is introduced, which meaning is shown by the following if-structures (169 – 179) and the relation Konr[`Prior,Sits[Priority]`] (180 – 183). The if-structure includes viality calculations of a description relation. The priority is determined by three exclusions. First, all theories which application quantities do not imply the according relation are excluded (169). Thence, theories with larger application quantities receive a higher priority, as the specialized theory elements are to be changed (173). Finally, the lower priority theories are analyzed after the actual viality of the description relation (176).

### 9.3.3 Konq[T,t] (270 – 1499)

Since the description of conclusion and guess making is crucial to understand the mechanisms of theory development, they are formulated explicitly as connection operators between the situative matrices and will be discussed in 9.3.3.1 and 9.3.3.2. Afterwards, the first three experiments are to be described precisely. Kont[Ex4], similarly to Konq[Ex4], will be discussed separately in 9.3.4.

### 9.3.3.1 Conclusion (294 – 331)

The conclusion process is described by Konw[1,2] (294 – 331). Before classification of the shown elements of Sits[R] (280 – 284), the organization into groups is formulated in line 296 by the mutual function identifying mutual attributes of finalities of the relevance situation. Thence, all elements are organized into groups with variable c. These groups are classified by application of the Classes function, as explained in 9.3.1 (319). The classified groups are described by the variable d. Each class consists of $\gamma(d)$ groups with variable c(d) (323). Afterwards, the conclusion situation is reformulated as a finality by applying the Möbius curvature in lines 327 and 328.

### 9.3.3.2 Guess Making (346 – 380)

After conclusion is done, Konw[2,3] (346 – 380) describes the formulation of guesses based on the classification. In this regard, w1 and w2 are different, arbitrarily chosen groups out of the classifications (348, 349). Thence, a relation between them is defined, which is afterwards interpreted by Konf[int] as the relation between a subject and an object (354 – 356). Hence, the guess can be formulated as Sits[G] (358 – 376), whereas the causal operator is defined as the classification of Konr[w1,w2] (357). Finally, the situation is also developed into a finality by Möbius curvature. The following situation, Sits[3], includes an according conclusion operator between the conclusion itself and the formulated guess (404 – 407).

### 9.3.3.3 KonF[Ex1] (423 – 581)

In the beginning of every experimental function, the experimental situation is declared. Since the first experiment is of type one, the experimental situation is the relevance situation from Kont[1]. The crucial condition is the viality of the description relation between guess and experimental situation (427 – 430). The integer Ex1A occupies the values zero and one, depending on whether the viality is omnivial or not (whether the guess explains the experimental situation properly or not). The integer Ex1loop is used as an indicator whether new guesses are able to be made. This notification is used by every experiment. As long as this is possible, the loop proceeds to create new guesses until either all possibilities run out or a successful explanation is achieved. While new guesses have to be made, each unsuccessful one is stored in a memory situation (500 – 522). If a description succeeds, the formation of model and application quantity is induced (440 – 486) and the algorithm proceeds to Kont[2], which formulates the theory element by rearranging the situative elements (625 – 669).

### 9.3.3.4 KonF[Ex2] (670 – 760)

As an experiment of type two, KonF[Ex2] contains more than one experimental situation, since it applies the basal theory element on experiments chosen according to its application quantity. Each time an experiment is explained by the theory, the according situative matrix is added into the relevance situation (680 – 687). Henceforth, this matrix can be regarded as a memory situation for all experiments which have been explained successfully. If new experiments are chosen, it is a crucial criterion that they differ from those which are already explained. Otherwise, the evolution of a theory cannot be guaranteed consequently. As long as the explanation succeeds, new experiments are chosen iteratively. However, as soon as an experiment is failed to be explained, the current theory is to be transformed into the core theory element and the crucial experiment is declared as the relevance situation, which is changed into Sits[R3] (722 – 737 if the failed explanation occurs after the first time, 742 – 757 if the first experiment is failed to be explained). In case of failed explanation, the program proceeds to build the first belt theory element in Kont[3] (783 – 983) by following the same scheme as Kont[1], KonF[Ex1] and Kont[2].

### 9.3.3.5 KonF[Ex3] (984– 1133)

Most of KonF[Ex3] is formulated similarly to KonF[Ex1]. As mentioned in 8.2.4 and 9.2.1, the crucial difference is given by proceeding directly to KonF[Ex4], if all possible guesses failed to explain the experimental situation and a modification is required (1130, 1131). Else, it proceeds to Kont[4] which rearranges the new and the basal theory element into a theory structure (1196 – 1281).

### 9.3.4 KonF[Ex4] and Konq[Ex4] (1282 – 1435, 1859 - 2011)

KonF[Ex4] and Konq[Ex4] represent the most important experiments. Indeed, they are the most complicated formulations as well. According to the principle mentioned in 8.1 and 8.2.5, belt elements are the first objects of modification (1305 – 1370, 1898 - 1944). Hence, as soon as the theory fails, the character task is set as Belt, which signalizes that belt modification is to be done in Konq[Rebuild], which is called afterwards (1328, for instance). If only one element is left in the belt, solely addition can be applied as modification. This case is implemented by another loop (1345 – 1368, 1919 - 1941), which sets `task` as `AdBelt` or `AdCore`, respectively. The maximum iteration steps are set by `l[2,max]` and `l[3,max]`. If the belt is erased completely and the maximum number of belt modifications is reached without providing a theory to explain the experiment, the same procedure starts for the core (1378 – 1426, 1951 - 1999). Similarly, the maximum iteration steps are set by `l[4,max]` and `l[5,max]`. Generally, this pattern is copied for `Konq[Ex4]`. One important difference between `KonF[Ex4]` and `Konq[Ex4]` is the recording of successful theories. Throughout `Konq[T,t]`, the actualization of a successful theory is done by `Kont[T]` (1436 – 1500). Accordingly, the notification `outcome` is set in `KonF[Ex4]` and other experiments (688, 713, 1310) in order to distinguish the origin of the current successful theory (1437 – 1452). In `Konq[Ex4]`, the actualization is done directly within the function (1881 – 1884, 2004 – 2007). Another difference is implied by the connection between `KonF[Ex3]` and `KonF[Ex4]` mentioned in 9.3.3.5. In order to differentiate the origin of the according theory, $ExA(3,4)$ is introduced before (1130, 1279). Dependent to its value, the translation of relevance and theory situation differs in `KonF[Ex4]` (1283 – 1303). In contrast, `Konq[Ex4]` requires solely one single translation (1862 – 1874).

### 9.3.5 Konq[Rebuild] (2012 – 4902)

Although the `Rebuild` sequence occupies the most part of PARTS, its structure is relatively simple. As has been depicted in 9.2.3, the three different modifications as well as an organization tensor are used for core and belt each. In the beginning of the sequence, the modification tensors are chosen according to the mentioned character `task`, as has been described in 9.3.4 (2013 – 2031). In general, the microstructure is quite similar to `Kont[1]`, `KonF[Ex1]` and `Kont[2]`. However, the differences are required to be presented in particular.

### 9.3.5.1 Kont[NiBelt] and Kont[NiCore] (2032 – 2130, 3512 – 3576)

The erasement tensors consist of two situative matrices each. In the first one, the priority relations are applied on the according situants (2063 – 2074, 3534 - 3537). The connecting operator (2076 – 2086, 3539 - 3549) simply erases the theory element with the lowest priority.

### 9.3.5.2 Kont[AdBelt] and Kont[AdCore] (2131 – 2780, 3577 – 4129)

The addition tensors are more complicated, since they are intended to elaborate a new theory element, following the same scheme as `Konq[T,t]`. Similarly to erasement tensors, the theory elements are ordered with respect to their priority first. Then, the element with lowest priority is separated (2184 – 2231, 3607 - 3615). Afterwards, the usual theory development starts, including an own experiment of type one (2232 – 2664, 3616 – 4041). In the next step, the new theory is combined with the separated one, which implies a specialization with respect to the experimental situation. Finally, the newly formed theory structure is included back into the according situant structure (2737 - 2779, 4100 – 4128). Similarly to the process in Konq[T,t], the iteration situations for guess making are formulated separately as tensors (2781 - 2864, 4130 – 4199).

### 9.3.5.3 Kont[ReBelt] and Kont[ReCore] (2865 – 3431, 4200 – 4685)

As depicted in 8.2.5 and 9.2.3, the replacement modification is formulated as a combination of erasement and addition, whereas there is no need to include the new theory with a low priority element. After the erasement of the element with lowest priority (2909 – 2919, 4230– 4240), the second step includes theory development and inclusion of the new theory directly into the according situant structure (2920 – 3430, 4241 – 4684). The iteration situations for guess making are formulated accordingly (3432 – 3511, 4686 – 4751).

### 9.3.5.4 Kont[ordBelt] and Kont[ordCore] (4752 – 4826, 4827 – 4901)

After more than one modification tensors have been applied on the theory structure, the organization tensors orders them with respect to the best description of the relevance situation. First, the final situations are included into a major situative matrix, whereas each situation includes a description relation and a function which counts the vialities (4754 – 4794, 4829 - 4869). Afterwards, the values are ordered and the situation with the highest value is chosen and declared as the outcome of modification process (4795 – 4823, 4870 – 4898).

### 9.3.6 Konq[Crosstest(tau1,tau2)] (1501 – 1858)

First, two different scientific branches Sith[T,tau1] and Sith[T,tau2] need to be recalled, whereas every commutating combination of a set has to be regarded (1502, 1503). Afterwards, the application quantities are combined and possible combinations of models are tested, as has been described in 8.2.6 (1504 – 1519). If no simple combination satisfies the requirement, the RRM starts to find a new model containing the mentioned subjects (1533 – 1856). As has been said, RRM solely focuses on model and causality operator research (1604 – 1680). In general, it is similar to the usual procedure.

### 9.3.7 Konq[TrendCalc] (245 – 270)

The calculation of set evolution trend is relatively simple. For every step *n*, the relation between the actual and the former $\theta$ as well the changing rate, if possible, is calculated (248, 249). Afterwards, Sits[Thetarel_History] records the actual evolution trend (256 – 269).

**Fig. 9.1**: Flow diagram of PARTS macrostructure. The variable t is defined from one to $\theta$, *failmax* is the highest possible number of failed attempts *f* to decrease the amount of theories.

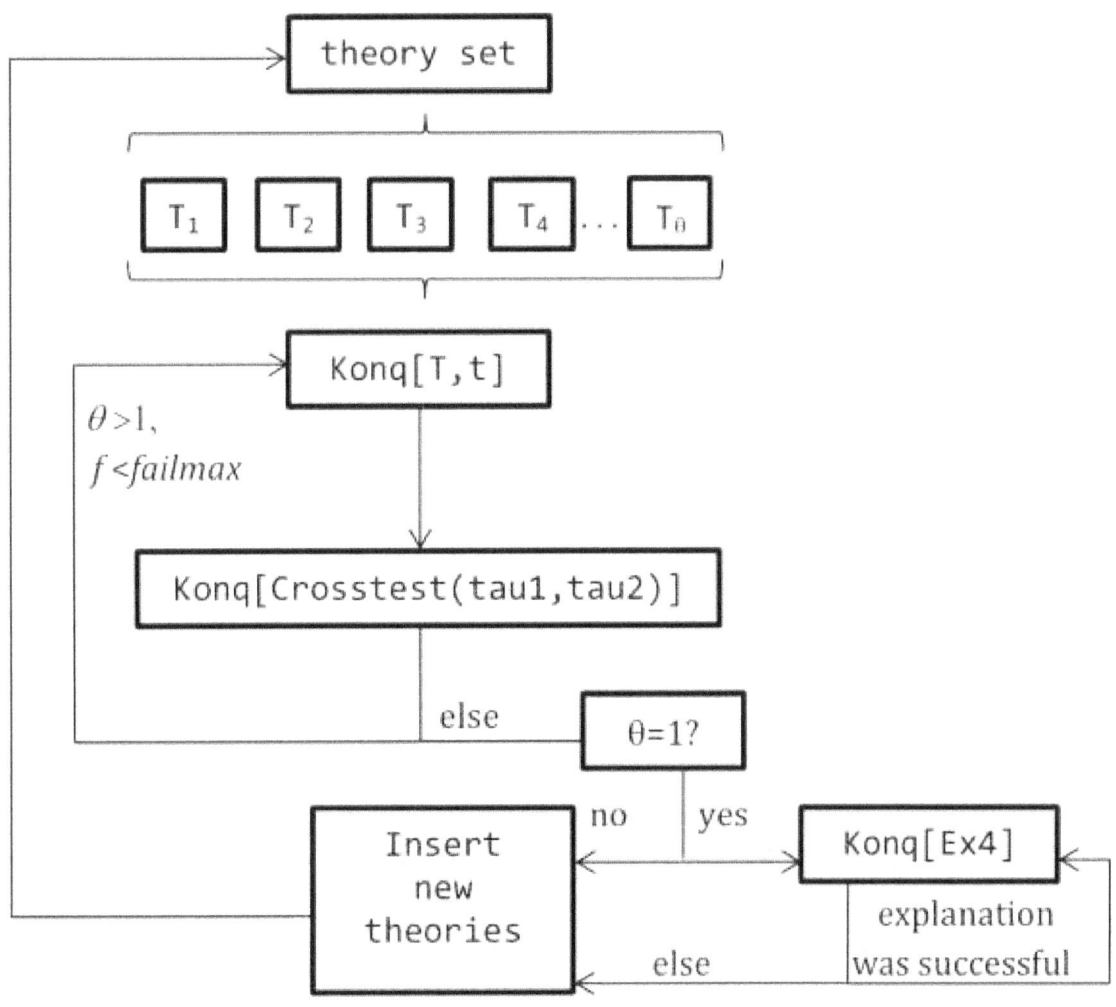

**Fig9.2**: Mesostructure of Konq[T,t]. The different tensors and functions have been discussed in 9.2.1.

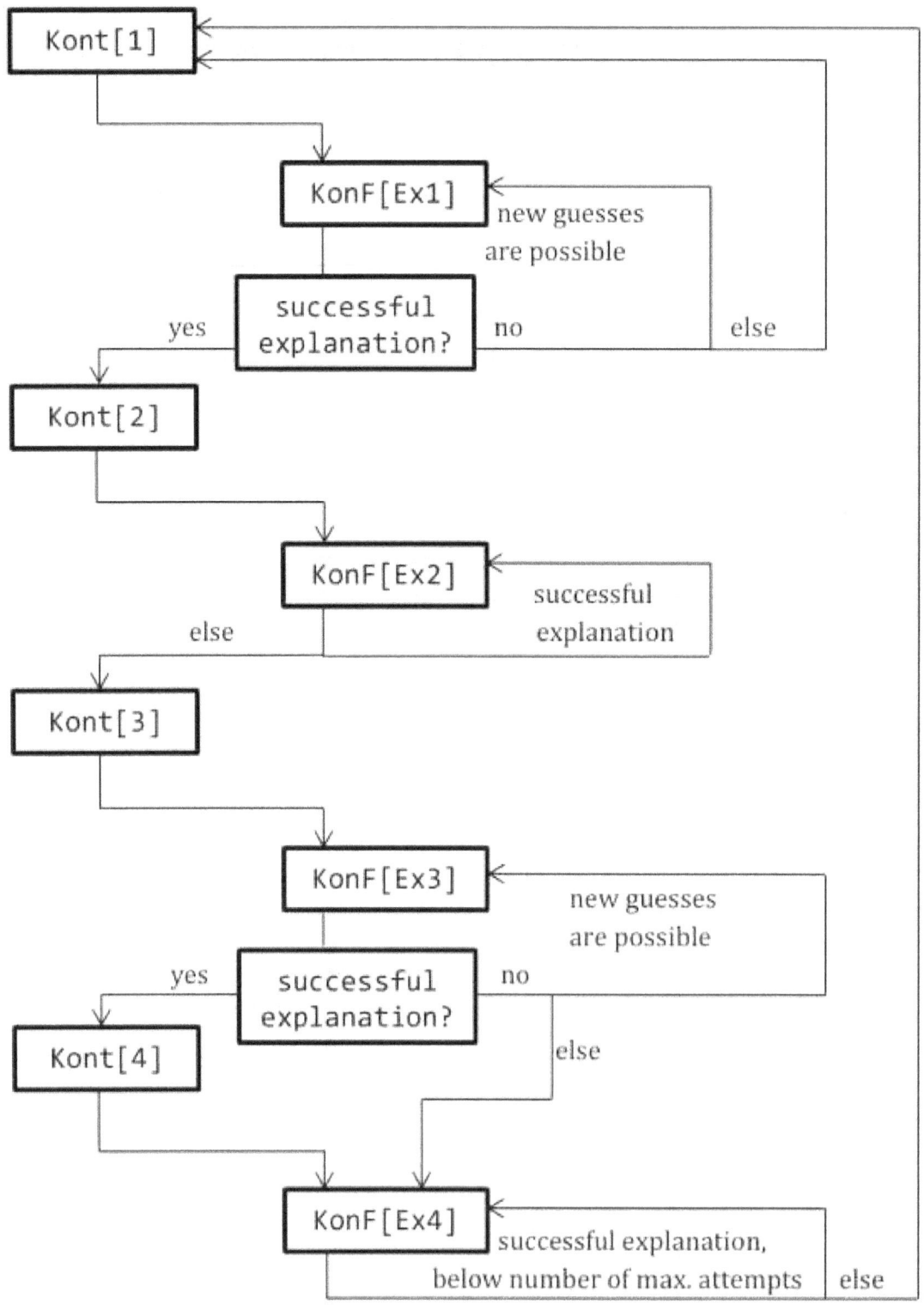

**Fig.9.3**: Mesostructure of Konq[Rebuild] as described in 9.2.3.

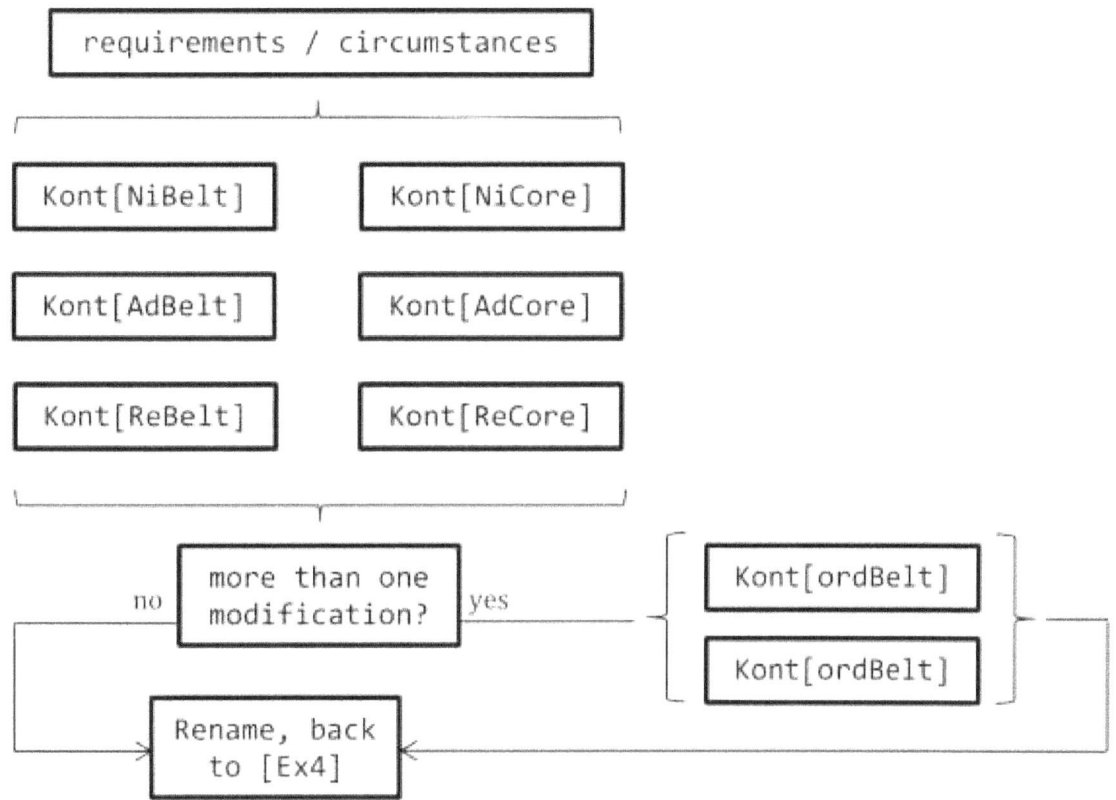

**Fig. 9.4**: Mesostructure of the different modification types: Ersament, Addition, Replacement as described in 9.2.3 and 9.3.4.

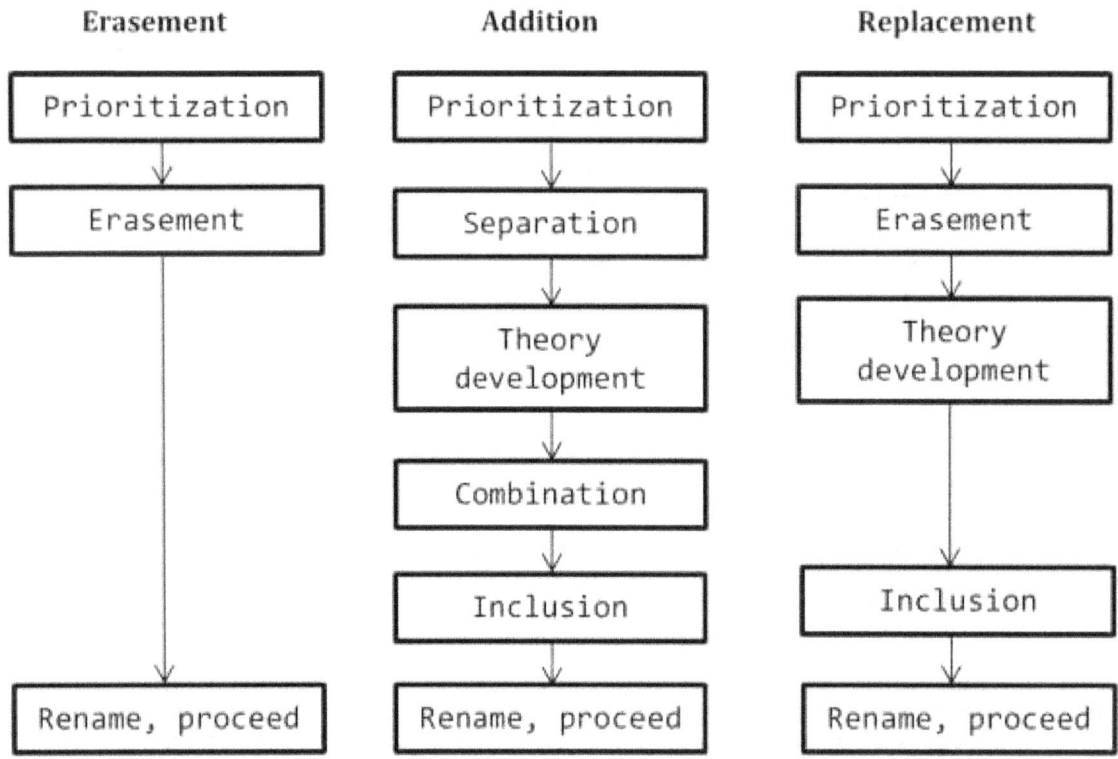

**Fig.9.5**: Mesostructure of Konq[Crosstest(tau1,tau2)]. Instead of $\tau_1$ or $\tau_2$, simply the numbers are chosen as indices.

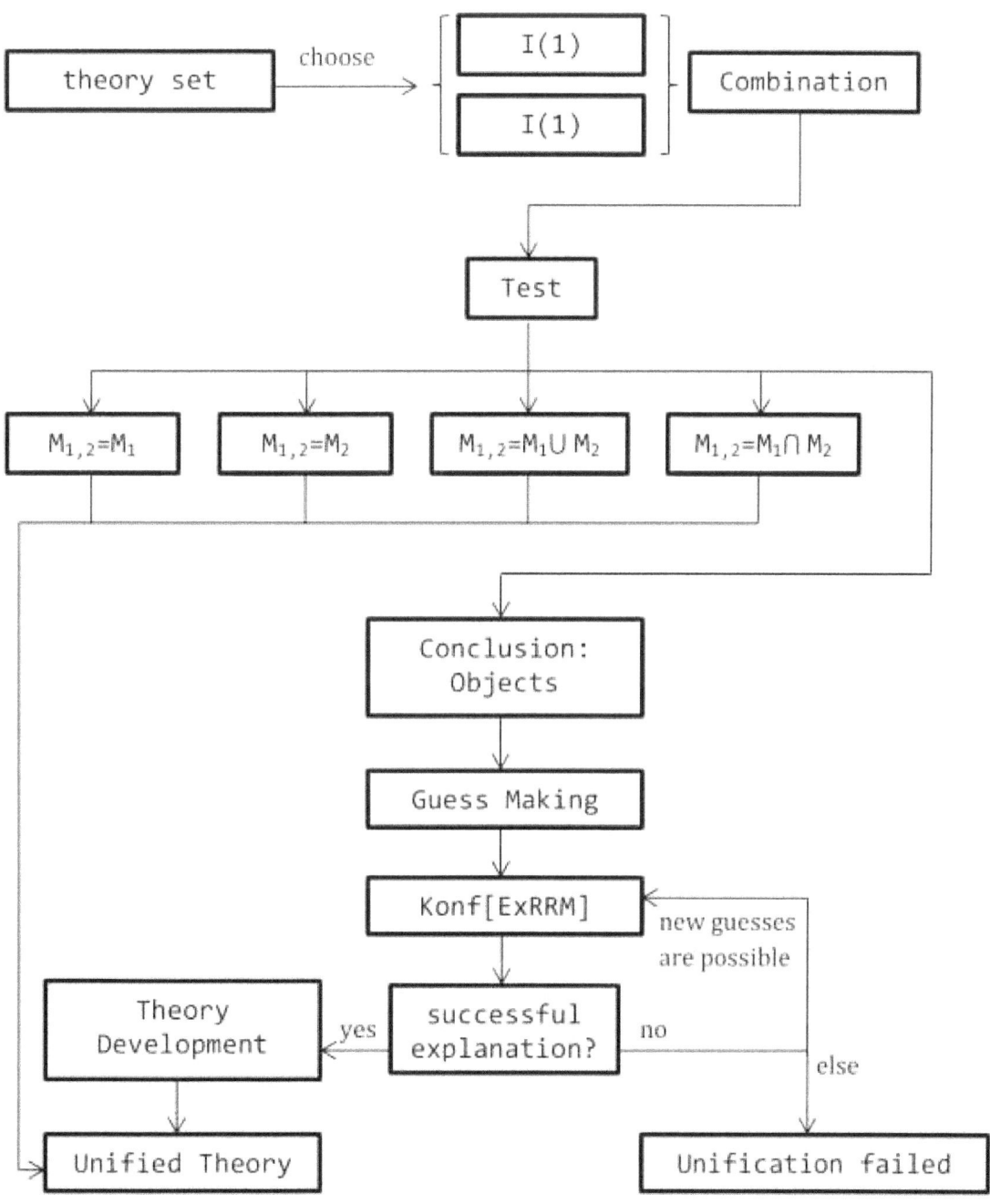

# 10. Conclusions and Remarks [7, 13, 17 – 19, 30 – 40]

## 10.1 General Remarks

### 10.1.1 Oversight and Insight

As has been shown in the first chapters, both the basal elements and precise instruments of an extraabstract language can be formulated or derived and explained from quite simple and generally accessible observations. Hence, one can derive an own language according to the own view. This means, that both MS and recomplex tensor sequence are solely *examples of an infinite set of languages*, in which none of them has a higher priority than the others. However, it is helpful to choose distinct groups of languages according to several factors, such as the own detection abilities, the required application or other circumstances. For instance, a being which is not capable of seeing, but solely communicates by voice, cannot use a language as the MS, since its syntax is shown visually. It needs a language which syntax is built by sound frequencies, for example. If there are intelligent beings which communicate solely by exchanging chemicals, their chosen language, consequently, would require an according syntax. Furthermore, if the application requires a language which can be understood by primates such as chimpanzees, who are not aware of metaabstract laws like us, we would choose a language which is restricted accordingly by their abstracticity grade on the first axis.

These thought experiments show that there are quite different and partially bizarre ways to construct syntax and basal building blocks of a language which satisfies the postulates of 1.3. We shall not consider any of them being superior or inferior with respect to other ones, as this would be a centralist perspective which theoretical basis is not provided by the postulates in any way.

PART 2, however, shows a transformation of MS, thus it is explicitly geared to it. Therefore, other sequence-like languages for computation must be evolved from other languages like MS, but not from the recomplex tensor sequence itself. Indeed, such a direct transformation is possible according to the postulates, but a translation would include two steps instead of one: First, we would need to transform the syntax of the matrices into a different syntax, thence we would need to transform the obtained sequential language into another one, including the instruments and sequential orders.

Hence, on the one hand, it is required to provide an orientation within these new theoretical dimensions. On the other hand, it is crucial to apply the pendant of the cosmological principle: For all languages, the properties of universe, therefore other languages as well, are the same, given sufficiently large copulae. Accordingly, there are no special perspectives, thus all possible perspectives are, objectively, equal to each other. By systematization, we may construct a system of orientation, but this does not affect the equality, even if we think it does.

## 10.1.2 Mutual Patterns of Dimensionality

As has been mentioned, the additional dimensions shown in chapter three are based on Kaluza's suggestion of additional space dimensions, which are coiled up in smaller scales. Such considerations led to development of the *string theory*, using *Calabi-Yau manifolds* as a theoretical basis for the multidimensional vibration of energy strings, causing the variety of subatomic particles. Accordingly, the additional dimensions of MS language in situant spaces can be regarded as a syntactical frame to reorganize the mathematical concepts which describe the auspicious, but complicated variations of the string theory. Generally, we see that the scale and the amount of dimensions are anti-proportional to each other in MS, leading to the assumption that there may be a more abstract reason for coiled extra dimensions, not solely affecting the physical world. In fact, the dimensionality changes with respect to the chosen perspective. By applying the formulae developed in 6.3, an according copula and thus a relation between attributes and experienced dimensions can be deduced. Furthermore, the classification of affects may be applied to find a verification strategy for the existence of strings, which exist in space scales smaller than the Planck length. Having these instruments, one can apply them on the theoretical concept of string theory and, probably, find a way to verify the theory experimentally.

## 10.2 Time, Causality and Symmetry Breaking

Beside quantum mechanics, which describes the physical world on small scales, general relativity describes physical phenomena on large scales, reaching up to cosmological scales. The basic concept of general (and special) relativity is the assumption that everything occurs within a four-dimensional fabric of space and time. At these scales, space occupies three dimensions, whereas the fourth dimension is occupied by time. It is clear that time in fact *is* one-dimensional, as we can solely move in one dimension in time – forwards and, at least theoretically, backwards. The attribute combinations described in 5.2 and 5.3 provide an approach to explain why time *must* be one-dimensional.

## 10.2.1 Interpretation of Time Flow in MS

The MS syntax does not allow a decent flow but a development of punctual situations. This view accords with quantum mechanical observations, which state that there is, indeed, a minimum detectable time-lapse, called the Planck time. If we choose a situation, it can evolve into a given quantity of other situations, or events, such as a ball resting on the top of a hill can roll down in different directions. In cosmology, this phenomenon is known as symmetry breaking and plays an important role in dynamics of the universe. We can apply the Möbius curvature to formulate the different events as finalities and sum over all phenomena obtaining a situative matrix. The initial event may be $Fin_A$, whereas all possible successor events may be called $Fin_1$ … $Fin_n$, where $n$ is an arbitrarily natural number. In order to sum up the possible successor events as a bulk of possibilities, we can include them into a situant space, whereas the meta finality may be called $Fin_B$ (10.1).

The summarization of all possible successors may sound arbitrarily and rather linguistically motivated, but since all these events have the common attribute of being a possible successor, a summarization does makes sense in a physical way as well. According to attribute combinations, there are different ways the attributes of these events can interact, depending on their nature with respect to combination properties (**chart 5.1**). Hence, due to their common attributes of being the possible successor, an interaction between these events takes place. Subsequently, every event's attributes change, leading to a differentiation between all possible successors.

## 10.2.2 Causality as Origin of Time

In order to formulate the idea depicted in 10.2.1 more precisely, we need to embed the concept into a physical interpretation. Since we intend to explain the nature of time, or time itself, we need to start off from another aspect of the physical world. The basic idea of this hypothesis states that causality is the reason why time exists, unlike the intuitive perspective. Accordingly, we exchange the time axis $t$ in the four-dimensional Cartesian system with the causality axis $K$. Since the y-axis is traditionally chosen to be the time axis which is now to be re-interpreted, this change can be called *Y-theorem*. As the x-axis remains the space axis summing up the three spatial dimensions, we can place the initial event $Fin_A$ in the origin of the Cartesian system, and the possible successor events on a half-circle around it, as they all have the same chance to be the actual successor of $Fin_A$. The equality of chances is represented by the same distance in 'space-causality' (**fig. 10.1**). Accordingly, the whole situation can be described by (10.1), whereas the shown affect can be interpreted as the transfer of the attribute of actuality.

Statistically, the combinatorial effects force the events to change their attributes, which affect their position in the Cartesian system (**fig. 10.2**). We may call the shortest distance between the initial event and the possible successors $l_{K,min}$.

**Fig. 10.1**: The space-causality Cartesian system (space=R, causality=K) with the initial event *Fin_A* in the origin, and all possible events in the surrounding, which means that all of them have the same distance to the initial event. Since the causality moves forward, the resulting figure is a half-circle.

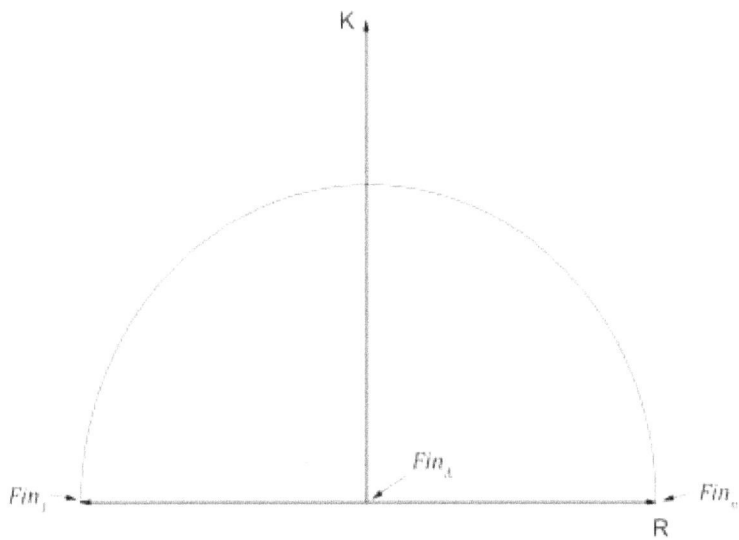

**Fig. 10.2**: Example for an event pattern after combinatorial effects. *Fin_j* is the actual successor of *Fin_A*.

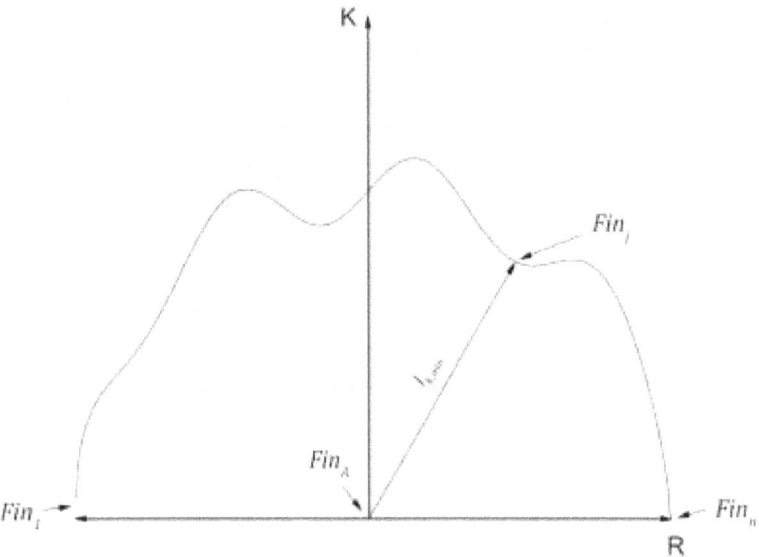

### 10.2.3 Derivation of Expression for Time according to the Y-Theorem

As we intend to find the actual successor of $Fin_A$, $Fin_j$, we need to find the finality with the highest compatibility after the combinatorial effects. Given the situation described by (10.1), we can formulate the compatibility requirement by (10.2).

$$(10.1) \quad Sit\sigma_I := \left\{ \begin{array}{l} Sit\varepsilon_A \\ Sit\eta_A \ Sit\eta_B : \left( \begin{array}{c} Sit\eta_1 \\ ... \\ Sit\eta_n \end{array} \right) \\ Fin_A \quad Fin_B : \left( \begin{array}{c} Fin_1 \\ ... \\ Fin_n \end{array} \right) \end{array} \right\}$$

$$(10.2) \quad C\left(Sit\varepsilon_A, Finx_j, Sit\eta_j\right) = \text{Æ}_{A,j} \cdot r\left(Sit\eta_j\right) = C_{\max}\left(Sit\varepsilon_A, Finx_B, Sit\eta_B\right)$$

with

$$C\left(Sit\varepsilon_A, Finx_B, Sit\eta_B\right) = \text{Æ}_{A,B} \cdot \sum_{i=0}^{n} r\left(Sit\eta_i\right)$$

The affect described in (10.1) solely includes the actual successor $Fin_j$, thus we can write:

$$(10.3) \qquad Kon\phi_1(Kon\rho_{A,B}) = \mu(Kon\rho_{A,B}) = Kon\rho_{A,j} := \langle Fin_A; Sit\,\eta_A; Fin_j \rangle = \Phi\chi(Sit\sigma_1, Fin_B)$$

We can interpret the combinatorial effects as an operator which arguments are the possible successor events. Accordingly, the actual successor is the outcome. In order to formulate the whole process more generally, we can sum up every operator for each actually occurring using the situation $Sit\sigma_T$:

$$(10.4) \qquad Sit\sigma_T := \left\{ \left| \left| \begin{matrix} Kon\varpi_{A,j} \\ ... \\ Kon\varpi_{j(m-1),j(m)} \end{matrix} \right| \right| \right\}$$

Every operator represents a single combination process occurring for each initial event. By applying the Möbius curvature on $Sit\sigma_T$, we obtain a finality we can call $Fin_T$:

$$(10.5) \qquad M_{C,3}^1(Sit\sigma_T) = Fin_T$$

In order to obtain a concrete formulation for the attribute of $Fin_T$, we write for the basal similarity grade used in (10.2):

$$(10.6) \qquad \mathcal{E}_{j(k-1),j(k)} = \frac{|E_{j(k-1)} \cap E_{j(k)}|}{|E_{j(k-1)}| + |E_{j(k)}|}$$

with

$$E_{j(k)} = \left\{ Sit\,\eta_{1(k)}, ..., Sit\,\eta_{n(k)} \right\}$$

The relevance factor can be formulated as the amount of entropy the system would gain if the according event is aimed. The reason for this decision is explained by the nature of the combinatorial effects and the relevance vector: As the possible combinations occur, it is more likely for the amount of initial information, represented by the attribute quantities, to decrease (**chart 5.1**). Decreasing information can be interpreted as increasing disorder, which is indeed an aspect of entropy, as it is depicted by the second law of thermodynamics. Thus, the increasing of disorder or loss of information is the relevance which must be as high as possible. We can write for the relevance factor:

$$(10.7) \qquad r(Sit\,\eta_x) = \Delta S(Fin_{x-1}, Fin_x) = S(Fin_x) - S(Fin_{x-1})$$

Together with the basal similarity grade, which is the highest for the nearest (space) and most similar (causality) event, these two factors choose the actual successor of $Fin_A$. As the space-causality distance between initial and successor event decreases with increasing compatibility and *vice versa*, we can formulate $l_{K,min}$ as the reciprocal value of compatibility:

$$(10.8) \qquad l_{K,\,min} = \left(C_{max}\left(Sit\varepsilon_{j(k-1)}, Finx_{j(k)}, Sit\eta_{j(k)}\right)\right)^{-1} = \left(\mathcal{E}_{j(k-1),j(k)} \cdot r\left(Sit\,\eta_{j(k)}\right)\right)^{-1}$$

By summing up the distance values for every operator shown in (10.4), we can deduce the punctually formulated time as a sum:

$$(10.9) \qquad Fin_T = \sum_{k=1}^{m} l_{K,\,min\,(k)}$$

If we observe macroscopic scales, we can transform this formula into an integral, whereas the integration variables are both space and causality:

$$(10.10) \qquad Fin_T = \iint l_{K,\,min\,(k)}\, dK\, dR$$

The shown derivation is solely an outline for the Y-theorem, showing a brief theoretical sketch of the concept of space-causality. Indeed, further investigations are required to verify the theoretical structure and concrete calculations. In the end, this approach has been done solely to explain the one-dimensional nature of time. Crucial verification criteria may be the consistency with observations and theoretical predictions under relativistic conditions. For example, time dilatation occurs in strong gravitational fields or at high speeds. Both means high quantities of energy. This leads as a bigger difference between initiative and successor events, which in turn lengthens $l_{K,min}$. By longer distances in space-causality, change rate decreases which ultimately leads to slower time flow with respect to systems with less energy.

## 10.3 Further Development of PARTS

The theoretical concept developed in chapter eight still has to be verified in actual scientific development, in order to develop both sequential concept and PARTS which into a more precise and sophisticated version. Presumably, the guess making, as well as the formulation of scientific laws (8.2.1) need to be specialized in order to obtain a self-operating paradigmatic algorithm. Accordingly, the essential function which develops the natural law operator, `Konf[Int]`, has been left unspecified within the algorithm. After the essential research is done, an explicit description of this function can be implemented without changing the rest of the source code redundantly. The same strategy has been applied on other functions, since their development would go beyond the scope of this work.

Another important point is that we may implement the possibility to interact with an operating system by suggesting own ideas, concepts and research results, which then would be applied on or included into current theory development. For instance, we may implement the Y-theorem concept and results of current experiments into an operating computer and thus verify whether this theorem actually helps to understand the observed physical phenomena. This human-machine interaction would combine the effectiveness of a solely algorithmic development and the manifold nature of solely anthropic evolution of scientific branches and thus lead to a highly efficient development of science itself. The realization of a crowd-based computational and anthropic scientific symbiosis can be realized using the World Wide Web. This would provide faster solutions to current scientific, sociological and economic problems. Therefore, the given PARTS version is solely the first design which is required to be developed by further research.

## 10.4 Consciousness and Artificial Intelligence

As introduced in 4.5, perspectives depict a finality's ability to be aware of its surroundings. Moreover, its copula potential can be described with respect to the finality's attributes, as has been shown in 6.3. We may formulate a situative matrix in which the observer is included as well. By finding the maximum copula potential for a situative matrix and solving equation (6.12) for the finality's attributes, we can deduce which attributes in fact enable a finality to be aware of itself. Subsequently, the chosen attributes can be described as the outcome of a combination process of situant attributes. These can be found by reverse application of combinatorial effects which is represented by **chart 5.1**. By choosing the right combination, one can formulate requirements for a being to develop self-awareness, or even consciousness. The identification process can be realized by applying PARTS on different possible explanations of consciousness development by combination. In fact, all possible combinations can be regarded as hypotheses, thus PARTS can apply the guess making and the first experiment of type one to find the required combination. This approach is not only helpful for decoding the processes inside the brain, but also for developing a new type of artificial intelligence.

Indeed, an AI which is aware of itself may be a threat to human beings, but it is our responsibility to program it to have what we refer to as 'common sense'. Thus, the consciousness system must be chosen to be grounded not solely on logics but also on moral standards. As an example, Lawrence Kohlberg's stages of moral development can be described in MS syntax, transferred into a tensor sequence and thus be implemented into the system which represents a basis for consciousness attributes. If the AI decides to ignore moral standards (or whatever other standards we implement as conditions), it ultimately stops to be conscious of itself.

Beside consciousness, a *strong* AI must not only mimic humanoid thinking but in fact do think like a human. In order to develop this property, we must elaborate a concept to include emotional, creative and social behavior as well. The reason why it is complicated to teach an AI to think like a human is given by its origin. While humans originated from animals by developing logical, analytical and numerical thinking, a strong AI must be developed from an AI 'species' which already includes this processing, but lacks of the other properties. Therefore, emotion, creativity and social behavior must be added to its currently possible abilities. In order to achieve this pendant of an evolution-like development, we can orientate ourselves on the *lateralization of brain function*, in which the left cerebral hemisphere is specialized for logical, linear and analytical cognitive processes, whereas the right hemisphere is specialized on emotional, creative, and timeless cognitive processes. Accordingly, we can evolve both processes separately. After development, we can create links between both algorithmic hemispheres. Given deduced consciousness attributes, this may lead to a holistic structure, in which consciousness results from specific processes in which both hemispheres are included.

In order to provide a blueprint for another important property we need to include, we may apply MS and tensor sequence on the well-known OCEAN model, in which personality attributes are detected by different types of behavior. These types can be interpreted as situant attributes after application of the Möbius curvature. By formulating the occurring combinatorial effects as tensor sequences, we can obtain an algorithm which provides an actual characteristic behavior pattern with respect to the AI's surroundings.

Alongside the briefly discussed properties, there are other important aspects which concepts need to be elaborated, before further research is to be undertaken. However, current development of quantum computers as well as ever-increasing sophistication of algorithms and efficiency of data storage will be helpful for this subject, regarding software and hardware design.

# References

**1)** Sagan, C. (1980): *Cosmos*, Random House, Inc.: New York.

**2)** Taleb, N. N. (2007): *The Black Swan: The Impact of the Highly Impropable*, Random House, New York.

**3)** Catt, J. A., Maslin, M. (2012): *The Prehistoric Human Time Scale*, The Geologic Time Scale 2012, Vol. 1, p. 1011 – 1032.

**4)** Jackendoff, R. (1983): *Semantics and cognition*, The MIT Press: Cambridge, Massachusetts: London.

**5)** Hart, J. A. (2010), *Globalization and Digitalization*, Transformacje, p. 227 – 237.

**6)** Kardashev, N. S. (1964): *Transmission of Information by Extraterrestrial Civilizations,* Soviet Astronomy, Vol. 8, p. 217 – 221.

**7)** Suntola, T. (2018): *The short history of science – or the long path on the union of metaphysics and empiricism. Third, complemented edition.* Physics Foundations Society – The Finnish Society for Natural Philosophy.

**8)** Mackey, M. C., Santillán, M. (2005): *Mathematics, Biology, Physics: Interactions and Interdependence*, Notices of the AMS, Vol. 52, 8, p. 832 – 840.

**9)** Quine, W. V. (1983): *16th World Congress of Philosophy. Volume 1*, p. 439 – 446.

**10)** Sankey, H. (1994): *The Incommensurability Thesis*, Avebury Press: Aldershot.

**11)** Bartels, A., Stöckler, M. (Ed.), (2007, 2009): *Science Theory. A Study Book. 2nd edition*. Mentis Verlag GmbH: Paderborn.

**12)** Brillouin, L. (1962, 2013): *Science and Information Theory. 2nd edition*, Academic Press, Inc., New York.

**13)** Newmeyer, F. J. (Ed.), (1988, 1989): *Linguistics: The Cambridge Survey. Volume I. Linguistic Theory: Foundations*, Cambridge University Press, Cambridge.

**14)** Beak, G. H., Snijman, D. A. (1978): *A botanical note – the Fibonacci series of numbers*, Veld & Flora – journals.co.za (01.06.2020).

**15)** Haldane, J. B. S. (1927): *Possible Worlds and Other Essays*, Chatto and Windus: London. Citation p. 286.

**16)** Davis Jr., G. A. (1946): *The origin of Ursa Major*, Popular Astronomy, Vol. 54.

**17)** Kaluza, T. (1921): *On the Unification Problem of Physics*, Preuss. Akad. Wiss. Berlin (Math. Phys.). p. 966 – 972.

**18)** Greene, B. (2006): *The elegant universe: Superstrings, Hidden Dimensions, and the Quest for the Ultimate Theory. 12th edition.* Wilhelm Goldmann Verlag: Munich.

**19)** Bieri, P. (2001): *Das Handwerk der Freiheit. Über die Entdeckung des eigenen Willens (The Craft of Freedom. On the Discovery of own Freedom),* Hanser: Munich.

**20)** Johnson, M. (1998): *Attribute-value logic and the theory of grammar*, Center for the Study of Language and Information: Stanford, CA.

**21)** Balzer, W., Moulines, C.U. & Sneed, J.D. (1987): *An Architectonic for Science*, Reidel: Dordrecht.

**22)** Gähde, U. (2002): *Holism, Underdetermination, and the Dynamics of Empirical Theories*, Synthese 130 (1), p. 69 – 90.

**23)** Keuth, H. (Ed.), (2004): *Karl Popper, The Logic of Scientific Discovery*, Reihe Klassiker Auslegen, Akademie-Verlag: Berlin.

**24)** Lakatos, I. (1970): *Falsification and Methodology of Scientific Research Programs*, Lakatos/Musgrave, p. 89 – 189.

**25)** Sneed, J.D. (1971, 1979): *The Logical Structure of Mathematical Physics*, Reidel: Dordrecht.

**26)** Bailer-Jones, D., Friebe, C. (2009): *Thomas Kuhn*, Mentis: Paderborn.

**27)** Feynman, R. (1965): *The Character of Physical Law*, MIT Press, Cambridge (Mass.).

**28)** Regionales Rechenzentrum für Niedersachsen / Leibnitz Universtität Hannover (2008): *Die Programmiersprache C. Ein Nachschlagwerk, 17th edition*, Zentralinstitut für angewandte Mathematik, Forschungszentrum Jülich GmbH: Hannover.

**29)** Heinold, B. (2012): *A Practical Introduction to Python Programming*, Department of Mathematics and Computer Science: Mount St. Mary's University.

**30)** Ellis, G., Lanza, A., Miller, J. (Ed.), (1993): *The Renaissance of General Relativity and Cosmology: A Survey to Celebrate the 65th Birthday of Dennis Sciama*, Cambridge University Press: New York.

**31)** Yau, S. T. (1978): *On The Ricci Curvature of a Compact Kähler Manifold and the Complex Monge-Ampère Equation, I*, Communications on Pure and Applied Mathematics, 31 (3), p. 339 – 411.

**32)** Baez, J. C., Fritz, T., Leinster, T. (2011): *A Characterization of Entropy in Terms of Information Loss*, Entropy, 13, p. 1945 – 1957.

**33)** Einstein, E. (1917): *Cosmological Considerations in the General Theory of Relativity*, Königlich Preusische Akademie der Wissenschaften: Berlin. Sitzungsberichte (1917), p. 142 – 152.

**34)** Kazanas, D. (1980): *Dynamics of the Universe and Spontaneous Symmetry Breaking*, Astrophysical Journal, Part 2 – Letters to the Editor, vol. 241, p. L59 – L63.

**35)** Velmans, M. (2009): *How to define Consciousness – and how not to define Consciousness*, Journal of Consciousness Studies, 16 (5): Goldsmiths, University of London, p. 139 – 156.

**36)** Barr, A., Feigenbaum, E. A. (Ed.), (1982): *The Handbook of Artificial Intelligence. Vol. 2*, William Kaufmann, Inc.: Los Altos, CA; HeurisTech Press: Stanford, CA.

**37)** Serov, A. (2014): *Subjective Reality and Strong Artificial Intelligence*, Machinery Research Institute, Russian Academy of Sciences, Moscow.

**38)** Arbib, M. A. (1975): *Artificial Intelligence and Brain Theory: Unities and Diversities*, Annals of Biomedical Engineering, 3, p. 238 – 274.

**39)** Ocklenburg, S., Güntürkün, O. (2018): *The Lateralized Brain: The Neuroscience and Evolution of Hemispheric Asymmetries*, Academic Press: London.

**40)** McCare, R. R., John, O. P. (1992): *An Introduction to the Five-Factor Model and its Applications*, Journal of Personality, 60 (2), p. 175 – 215.

# PART 3:
# PARTS SOURCE CODE

## Foreword to PARTS

The following source code of PARTS has been elaborated according to the process depicted in PART 2. The line numbers are shown for orientation and agree with the declarations in chapter nine. It is free to change the source code, as it solely represents, like the whole work, a suggestion.

It is important to note that although I tried to optimize the formulation, there is no guarantee that the source code is completely faultless. I am thankful for everyone who alludes to found mistakes.

Sincerely,

Waldemar Schwarzkopf

```
1    >Konq[PARTS]
2    {>Konq[Application]
3      >character:
4      >NAME, task, chi
5      >reaL:
6      >outcome, thetarel(n), thetarate(n)
7      >integer:
8      >theta, thetaold, thetanew, tau1, tau1, failmax, n, nrun, nrate, Ct(tau1,tau2)
9      >finish(X1), finish(X2), rel1, rel2,
10     >alpha, alphanew, beta, betanew, gamma, delta
11     >m, delta(Subject), delta(Object), gamma(d(Subject)), gamma(d(Object))
12     >alphaA, alphaAnew, betaA, betaAnew, gammaA,  deltaA
13     >qc, deltaA(SubjectA), deltaA(ObjectA), gammaA(dA(SubjectA)), gammaA(dA(ObjectA))
14     >Ex1A, Ex1loop, Ex2A, Ex2loop, Ex3A, Ex3loop,
15     >Ex40, Ex4A, Ex4B, Ex4C, Ex4D, Ex4E,
16     >l[0], l[1], l[2], l[3], l[4], l[5], >NiB, AdB, ReB, NiC, AdC, ReC
17     >i, v(I), v(M), j, p(Io), p(Mo), p2(Io), p2(Mo), q(Ii), q(Ii), q2(Ii), q2(Mi), iA
18     >ExRRMA, i_RRM, i_RRMmax, ExRRMloop,
19     >i_AdB, ExAdBloop, ExAdBA, i_ReB, ExReBloop, ExReBA, i_AdC, ExAdCoop, ExAdCA, i_ReC,
20     >ExReCloop, ExReCA,
21     >alpha_AdB, alphanew_AdB, beta_AdB, betanew_AdB, gamma_AdB, delta_AdB,
22     >m_AdB, delta_AdB(Subject_AdB), delta_AdB(Object_AdB), gamma_AdB(d_AdB(Subject_AdB)),
23     >gamma_AdB(d_AdB(Object_AdB))
24     >alpha_ReB, alphanew_ReB, beta_ReB, betanew_ReB, gamma_ReB, delta_ReB,
25     >m_ReB, delta_ReB(Subject_ReB), delta_ReB(Object_ReB), gamma_ReB(d_ReB(Subject_ReB)),
26     >gamma_ReB(d_ReB(Object_ReB))
27     >alpha_AdC, alphanew_AdC, beta_AdC, betanew_AdC, gamma_AdC, delta_AdC,
28     >m_AdC, delta_AdC(Subject_AdC), delta_AdC(Object_AdC), gamma_AdC(d_AdC(Subject_AdC)),
29     >gamma_AdC(d_AdC(Object_AdC))
30     >alpha_ReC, alphanew_ReC, beta_ReC, betanew_ReC, gamma_ReC, delta_ReC,
31     >m_ReC, delta_ReC(Subject_ReC), delta_ReC(Object_ReC), gamma_ReC(d_ReC(Subject_ReC)),
32     >gamma_ReC(d_ReC(Object_ReC))
33    >print: „integer input: l[2,max], l[3,max], l[4,max], l[5,max])"
34    >input (l[2,max], l[3,max], l[4,max], l[5,max])
35    >variable:
36    >t[1:theta]
37    >t(plus)[2:theta]
38    >t(diff)[thetaold+1:theta]
39    >t(exp)[thetaold+1:theta+failmax]
40    >fail[1:failmax]
41    >SPACE[Ч,З,С,Ц]
42    >e[SPACE](NAME)
43    >counter(X1)[1:finish(X1)]
44    >counter(X2)[1:finish(X2)]
45    >a[1:alpha]
46    >anew[1:alphanew]
47    >b(a)[1:beta(a)]
48    >bnew(anew)[1:betanew(anew)]
49    >c[1:gamma]
50    >c(d)[1:gamma(d)]
51    >d[1:delta]
52    >w[1:m]
53    >d(Subject)[1:delta(Subject)]
54    >d(Object)[1:delta(Object)]
55    >c(d(Subject))[1:gamma(d(Subject))]
56    >c(d(Object))[1:gamma(d(Object))]
57    >u(I)[1:v(I)]
58    >u(M)[1:v(M)]
59    >x(Io)[1:p]
60    >x(Mo)[1:p(Mo)]
```

```
61    >x2(Io)[1:p2(Io)]
62    >x2(Mo)[1:p2(Mo)]
63    >aA[1:alphaA]
64    >aAnew[1:alphaAnew]
65    >bA(aA)[1:beta(aA)]
66    >bAnew(aAnew)[1:betaAnew(aAnew)]
67    >cA[1:gammaA]
68    >cA(dA)[1:gammaA(dA)]
69    >dA[1:deltaA]
70    >yc[1:qc]
71    >dA(SubjectA)[1:deltaA(SubjectA)]
72    >dA(ObjectA)[1:deltaA(ObjectA)]
73    >cA(dA(SubjectA))[1:gammaA(dA(SubjectA))]
74    >cA(dA(ObjectA))[1:gammaA(dA(ObjectA))]
75    >y(Ii)[1:q(Ii)]
76    >y(Mi)[1:q(Mi)]
77    >y2(Ii)[1:q(Ii)]
78    >y2(Mi)[1:q(Mi)]
79    >a_AdB[1:alpha_AdB]
80    >anew_AdB[1:alphanew_AdB]
81    >b_AdB(a_AdB)[1:beta_AdB(a_AdB)]
82    >bnew_AdB(anew_AdB)[1:betanew_AdB(anew_AdB)]
83    >c_AdB[1:gamma_AdB]
84    >c_AdB(d_AdB)[1:gamma_AdB(d_AdB)]
85    >d_AdB[1:delta_AdB]
86    >w_AdB[1:m_AdB]
87    >d_AdB(Subject_AdB)[1:delta_AdB(Subject_AdB)]
88    >d_AdB(Object_AdB)[1:delta_AdB(Object_AdB)]
89    >c_AdB(d_AdB(Subject_AdB))[1:gamma_AdB(d_AdB(Subject_AdB))]
90    >c_AdB(d_AdB(Object_AdB))[1:gamma_AdB(d_AdB(Object_AdB))]
91    >a_ReB[1:alpha_ReB]
92    >anew_ReB[1:alphanew_ReB]
93    >b_ReB(a_ReB)[1:beta_ReB(a_ReB)]
94    >bnew_ReB(anew_ReB)[1:betanew_ReB(anew_ReB)]
95    >c_ReB[1:gamma_ReB]
96    >c_ReB(d_ReB)[1:gamma_ReB(d_ReB)]
97    >d_ReB[1:delta_ReB]
98    >w_ReB[1:m_ReB]
99    >d_ReB(Subject_ReB)[1:delta_ReB(Subject_ReB)]
100   >d_ReB(Object_ReB)[1:delta_ReB(Object_ReB)]
101   >c_ReB(d_ReB(Subject_ReB))[1:gamma_ReB(d_ReB(Subject_ReB))]
102   >c_ReB(d_ReB(Object_ReB))[1:gamma_ReB(d_ReB(Object_ReB))]
103   >a_AdC[1:alpha_AdC]
104   >anew_AdC[1:alphanew_AdC]
105   >b_AdC(a_AdC)[1:beta_AdC(a_AdC)]
106   >bnew_AdC(anew_AdC)[1:betanew_AdC(anew_AdC)]
107   >c_AdC[1:gamma_AdC]
108   >c_AdC(d_AdC)[1:gamma_AdC(d_AdC)]
109   >d_AdC[1:delta_AdC]
110   >w_AdC[1:m_AdC]
111   >d_AdC(Subject_AdC)[1:delta_AdC(Subject_AdC)]
112   >d_AdC(Object_AdC)[1:delta_AdC(Object_AdC)]
113   >c_AdC(d_AdC(Subject_AdC))[1:gamma_AdC(d_AdC(Subject_AdC))]
114   >c_AdC(d_AdC(Object_AdC))[1:gamma_AdC(d_AdC(Object_AdC))]
115   >a_ReC[1:alpha_ReC]
116   >anew_ReC[1:alphanew_ReC]
117   >b_ReC(a_ReC)[1:beta_ReC(a_ReC)]
118   >bnew_ReC(anew_ReC)[1:betanew_ReC(anew_ReC)]
119   >c_ReC[1:gamma_ReC]
120   >c_ReC(d_ReC)[1:gamma_ReC(d_ReC)]
```

```
121    >d_ReC[1:delta_ReC]
122    >w_ReC[1:m_ReC]
123    >d_ReC(Subject_ReC)[1:delta_ReC(Subject_ReC)]
124    >d_ReC(Object_ReC)[1:delta_ReC(Object_ReC)]
125    >c_ReC(d_ReC(Subject_ReC))[1:gamma_ReC(d_ReC(Subject_ReC))]
126    >c_ReC(d_ReC(Object_ReC))[1:gamma_ReC(d_ReC(Object_ReC))]
127        >Kont[Entanglement]
128            >Sits[Entanglement]
129             >structure
130                    >Site[Effect]
131                    >Sits[B]; Fin[A]
132             >structure end
133             >Konr[EffectA]
134              >Site[Effect]
135              >Fin[A]
136             >Konr[EffectA]end
137             >Konr[EffectB]
138              >Site[Effect]
139              >Sith[B]
140             >Konr[EffectB]end
141             >Konf[Entanglement]
142              >of Konr[EffectA]
143             >Konr[EffectB]
144             >Konf[Entanglement]end
145            >Sits[Entanglement]end
146    >Kont[Entanglement]end
147    >Kont[Priority]
148     >counter(I)[1:abs(Sith[I])]
149     >counter(M)[1:abs(Fin[M])]
150     >counter(A)[1:abs(column(Sit[A]))]
151     >coutner(M,p)[1:abs(Fin[M])]
152     >1<=pa,pb<=abs(Fin[M])
153     >pa\=pb
154            >Sits[Priority]
155             >structure
156                    >Sits[A]
157                     >structure
158                            >Sith[A,counter(A)]; Fin[A,coutner(A)]
159                     >structure end
160                    >Sits[A]end
161                    >Sith[I,coutner(I)]; Fin[M,counter(M)]
162             >strucutre end
163             >Konr[Descript(M,A)]
164              >Sith[I,coutner(I)]; Fin[M,coutner(M)]
165              >E[Ц,e[Ц](A)]
166             >Konr[Descript(M,A)]end
167            >Sits[Priority]end
168     >counter(M)(p)[1:abs(Fin(M))]
169     >if ((Sith[A,counter(A)] element of (Sith[I,counter(I)(p)]) and (Sith[A,pb] not
170     >element of (Sith[I,counter(I)(p)])):
171      >pa>pb
172     >endif
173     >if (abs(Sith[I,pa])<abs(Sith[I,pb])):
174      >pa>pb
175     >endif
176     >if (((Sith[A,pa] and Siht[A,pb]) element of (Sith[I,counter(I)(p)])) and
177     >(abs(viality(Konr[Descript((M,pa),A]))>abs(viality(Konr[Descript((M,pb),A])))):
178      >pa<pb
179     >endif
180     >Konr[Prior,Sits[Priority]]
```

```
181          >Fin[M,counter(M,p)]
182          >Fin[M,counter(M,p)+1]
183         >Konr[Prior,Sits[Priority]]end
184        >Kont[Priority]end
185    >Konq[Application]end}
186   >KonF[Application]
187    >of Konq[Application]
188    >apply on ALL Konq
189   >KonF[Application]end
190   {>Konq[Paradigmatic algorithm]
191     >tau1\=tau2
192     >1<=tau1,tau2<=theta
193     >input:
194      >Sits[R,t]
195       >E[Ц,e[Ц](R,t)]
196      >Sits[R,tau1]\=Sits[R,tau2]
197       >E[Ц,e[Ц](R,tau1)]\=E[Ц,e[Ц](R,tau2)]
198     >while (0=0):
199      >fail=0
200      >n=0
201      >while ((theta>1) and (fail<failmax)):
202       >Konq[T,t]
203       >thetaold=theta
204       >repeat (ALL tau1,tau2)
205        >Konq[Crosstest(tau1,tau2)]
206        >if (Ct(tau1,tau2)=1):
207         >thetanew=theta
208         >n=n+1
209         >Konq[TrendCalc]
210         >fail=0
211        >elif (Ct(tau1,tau2)=0):
212         >thetanew=theta
213         >n=n+1
214         >Konq[TrendCalc]
215         >fail=fail+1
216        >endif
217       >repeatend
218      >else
219       >if (theta=1):
220        >thetanew=theta
221        >n=n+1
222        >Konq[TrendCalc]
223        >Konq[Ex4]
224        >input:
225         >Sits[R,t(plus)]
226          >E[Ц,e[Ц](R,t(plus))]
227       >else
228        >n=n+1
229        >Konq[TrendCalc]
230        >thetanew=theta
231        >if (thetanew<thetaold):
232         >input:
233          >Sits[R,t(diff)]
234           >E[Ц,e[Ц](R,t(diff))]
235        >elif (thetanew=thetaold)
236         >input:
237          >Sits[R,t(exp)]
238           >E[Ц,e[Ц](R,t(exp))]
239        >endif
240       >endif
```

```
241        >endloop
242      >endloop
243     >Konq[Paradigmatic Algorithm]end}
244    {>Konq[TrendCalc]
245      >thetanew(n)=thetanew
246      >thetaold(n)=thetaold
247      >thetarel(n)=thetanew(n)/thetaold(n)
248      >thetarate(n)=(thetarel(n+1)-thetarel(n-1))*0.5
249     >nrun[1:n]
250     >nrate[2:n-1]
251     >Fin[Thetarel,nrun]=nrun
252     >Fin[Thetarate,nrate]=nrate
253     >Sith[Thetarel,nrun]=thetarel(nrun)
254     >Sith[Thetarate,nrate]=thetarate(nrate)
255              >Sits[Thetarel_History]
256               >structure
257                    >Sith[Thetarel,nrun]; Fin[Thetarel,nrun]
258                    >Sith[Thetarate,nrate]; Fin[Thetarate,nrate]
259               >structure end
260               >Konf[Rel,Rate]
261                >of Е[Ц,е[Ц](Thetarel_History)]
262                >ALL nrate=nrel
263                >Konr[Rel,Rate]
264                 >Sith[Thetarel,nrun]; Fin[Therarel,nrun]
265                 >Sith[Thetarate,nrate]; Fin[Thetarate,nrate]
266                >Konr[Rel,Rate]end
267               >Konf[Rel,Rate]
268              >Sits[Thetarel_History]end
269     >Konq[TrendCalc]end}
270    {>Konq[T,t]
271          >Kont[0]
272               >Sits[R]=Sits[R,t]
273                >Е[Ц,е[Ц](R)]=Е[Ц,е[Ц](R,t)]
274          >Kont[0]end
275          >Kont[1]
276               >Sits[1]
277                >structure
278                 >Sits[R]
279                  >structure
280                     >Sith[R]; Fin[R]
281                            >Sith[R,a]; Fin[R,a]
282                                   >Sith[R,a,b(a)]
283                            >Sith[R,a]; Fin[R,a] end
284                     >Sith[R]; Fin[R] end
285                  >strucutre end
286                 >Sit[R] end
287                     >Sith[Obs1]; Fin[Obs]
288                >structure end
289                >Konp[Obs1]
290                 >of Sith[Obs1], Fin[Obs]
291                 >Е[Ц,е[Ц](1)], Е[Ц,е[Ц](R)]
292                >Konp[Obs1]end
293               >Sits[1]end
294          >Konw[1,2]
295           >of Sits[1]
296           >Sith[C1,c]=mutual(Sith[R,a,b(a)])
297           >while (c=0):
298             >\Sits[R]
299            >Sits[Rnew]\=Sits[R]
300                 >Sits[Rnew]
```

5

```
301              >structure
302                  >Sith[Rnew]; Fin[Rnew]
303                      >Sith[Rnew,anew]; Fin[Rnew,anew]
304                          >Sith[Rnew,anew,bnew(anew)]
305                      >Sith[Rnew,anew]; Fin[Rnew,anew] end
306                  >Sith[Rnew]; Fin[Rnew] end
307              >strucutre end
308          >Sits[Rnew] end
309      >Sits[R]=Sits[Rnew]
310       >E[Ц,e[Ц](R)=E[Ц,e[Ц](Rnew)]
311      >else
312              >Sits[C1]
313               >structure
314                  >Sith[C1]
315                      >Sith[C1,c]
316                  >Sith[C1] end
317              >structure end
318          >Sits[C1] end
319      >Sith[d]=Classes(Sith[C1])
320              >Sits[C2]
321               >structure
322                  >Sith[d]
323                      >Sith[d,c(d)]
324                  >Sith[d] end
325              >structure end
326          >Sits[C2] end
327      >Sith[C], Sith[C,w]=Konw[M](Sits[C2])
328       >E[Ч,e[Ч](C)]=E[3,e[3](C2)]
329      >Sits[2]
330     >endloop
331  >Konw[1,2]end
332          >Sits[2]
333           >structure
334           >Sits[R]
335            >E[Ц,e[Ц](R)]
336              >Sith[Obs2]; Fin[Obs]
337              >Sith[C]
338                  >Sith[C,w]
339              >Sith[C] end
340           >structure end
341           >Konp[Obs2]
342            >of Sith[Obs2], Fin[Obs]
343            >E[Ц,e[Ц](2)], E[Ц,e[Ц](R)]
344           >Konp[Obs2]end
345          >Sits[2]end
346  >Konw[2,3]
347   >of Sits[2]
348    >w1,w2=groups(m)
349   >w1\=w2
350   >Konr[w1,w2]
351     >Sith[C,w1]
352     >Sith[C,w2]
353    >Konr[w1,w2]end
354    >Sits[G]=Konf[int](Konr[w1,w2])
355    >E[Ч,e[Ч](Subject)]=Sith[C,w1]
356     >E[Ч,e[Ч](Object)]=Sith[C,w2]
357    >Konw[Causality]=Class(Konr[w1,w2])
358          >Sits[G]
359           >structure
360               >Sith[Subject]
```

```
361                              >Sith[Subject,d(Subject)]
362                                      >Sith[Subject,d(Subject),c(d(Subject))]
363                              >Sith[Subject,d(Subject)] end
364                      >Sith[Subject] end
365                      >Sith[Object]
366                              >Sith[Object,d(Object)]
367                                      >Sith[Object,d(Object),c(d(Object))]
368                              >Sith[Object,d(Object)] end
369                      >Sith[Object] end
370              >structure end
371              >Konw[Causality]
372               >of Sith[Subject], Sith[Subject,d(Subject)],
373               >Sith[Subject,d(Subject),c(d(Subject))]
374               >Sith[Object], Sith[Object,d(Object)], Sith[Object,d(Object),c(d(Object))]
375              >Konw[Causality]end
376          >Sits[G] end
377     >Sith[G]=Konw[M](Sits[G])
378      >E[Ч,e[Ч](G)]=E[З,e[З](G)]
379     >Sits[3]
380  >Konw[2,3]end
381              >Sits[3]
382               >structure
383               >Sits[R]
384                >E[Ц,e[Ц](R)]
385               >Sits[G]
386                >E[Ц,e[Ц](G)]
387                      >Sith[Obs3]; Fin[Obs]
388                      >Sith[C]
389                              >Sith[C,w]
390                      >Sith[C] end
391                      >Sith[G]
392                              >Sith[Subject]
393                                      >Sith[Subject,d(Subject)]
394                                              >Sith[Subject,d(Subject),c(d(Subject))]
395                                      >Sith[Subject,d(Subject)] end
396                              >Sith[Subject] end
397                              >Sith[Object]
398                                      >Sith[Object,d(Object)]
399                                              >Sith[Object,d(Object),c(d(Object))]
400                                      >Sith[Object,d(Object)] end
401                              >Sith[Object] end
402                      >Sits[G] end
403              >structure end
404              >Konw[Conc(C,G)]
405               >E[Ч,e[Ч](C)]
406               >E[Ч,e[Ч](G)]
407              >Konw[Conc(C,G)]end
408              >Konw[Causality]
409               >of Sith[Subject], Sith[Subject,d(Subject)],
410               >Sith[Subject,d(Subject),c(d(Subject))]
411               >Sith[Object], Sith[Object,d(Object)], Sith[Object,d(Object),c(d(Object))]
412              >Konw[Causality]end
413              >Konp[Obs3]
414               >of Sith[Obs3], Fin[Obs3]
415               >E[Ц,e[Ц](3)], E[Ц,e[Ц](R)]
416              >Konp[Obs3]end
417              >Konr[Descript(G,R)]
418               >E[Ч,e[Ч](G)]
419               >E[Ц,e[Ц](R)]
420              >Konr[Descript(G,R)]end
```

```
421          >Sits[3]end
422     >Kont[1]end
423     >KonF[Ex1]
424      >of Kont[1], Sits[3]
425      >Sits[Ex1]=Sits[R]
426       >E[Ц,e[Ц](Ex1)]=E[Ц,e[Ц](R)]
427      >Konr[Descript(G,Ex1)]
428       >E[Ч,e[Ч](G)]
429       >E[Ц,e[Ц](Ex1)]
430      >Konr[Descript(G,Ex1)]end
431     >if (Konr[Descript(G,Ex1)]=omnivial):
432      >Ex1A=1
433     >else:
434      >Ex1A=0
435     >endif
436     >Ex1loop=1
437     >i=1
438     >while (Ex1loop=1):
439      >if (Ex1A=1):
440       >Sits[R2]=Sits[Ex1]
441        >E[Ц,e[Ц](R2)]=E[Ц,e[Ц](Ex1)]
442          >Sits[Model]
443           >structure
444               >Sith[chi]; Fin[chi]
445                    >Sith[Subject]
446                         >Sith[Subject,d(Subject)]
447                             >Sith[Subject,d(Subject),c(d(Subject))]
448                         >Sith[Subject,d(Subject)] end
449                    >Sith[Subject] end
450                    >Sith[Object]
451                         >Sith[Object,d(Object)]
452                             >Sith[Object,d(Object),c(d(Object))]
453                         >Sith[Object,d(Object)] end
454                    >Sith[Object] end
455               >Sith[chi]; Fin[chi] end
456           >structure end
457           >Konw[NG]
458            >of Sith[Subject], Sith[Subject,d(Subject)],
459            >Sith[Subject,d(Subject),c(d(Subject))]
460            >Sith[Object], Sith[Object,d(Object)], Sith[Object,d(Object),c(d(Object))]
461           >Konw[NG] end
462          >Sits[Model] end
463          >Sits[h,I]
464           >structure
465               >Sith[chi]; Fin[chi]
466                    >Sith[Subject]
467                         >Sith[Subject,d(Subject)]
468                             >Sith[Subject,d(Subject),c(d(Subject))]
469                         >Sith[Subject,d(Subject)] end
470                    >Sith[Subject] end
471               >Sith[chi]; Fin[chi] end
472           >structure end
473          >Sits[h,I] end
474       >Class(Sith[chi],Fin[chi](Sits[h,I]))=Konf[M](E[Ц,e[Ц](Model)],Konw[NG])
475       >Konr[Subject,chi]
476         >Sith[Subject]
477         >Sith[chi]
478        >Konr[Subject,chi]end
479       >Konr[Subject,chi]=Konf[M](Konw[NG])
480       >Class(Konr[Subject,chi])=Ef
```

8

```
481        >Sith[I], Sith[I,u(I)]
482         >E[Ч,e[Ч](I)]=Class(Sith[chi],Fin[chi])
483        >Fin[M], Fin[M,u(M)]=Konw[M](Sits[Model])
484         >E[Ч,e[Ч](M)]=E[3,e[3](Model)]
485        >v(I)=abs(Class(Sith[chi],Fin[chi]))
486        >v(M)=abs(ALL(E[3,e[3](Model)]))
487        >Kont[2]
488      >elif (Ex1A=0):
489        >s[1:i]
490        >w1(i)=w1
491        >w2(i)=w2
492        >Sith[i,Subject(i),d(Subject(i)),
493        >c(d(Subject(i)))]=Sith[Subject,d(Subject),c(d(Subject))]
494        >Sith[i,Subject,d(Subject(i))]=Sith[Subject,d(Subject)]
495        >Sith[i,Subject(i)]=Sith[Subject]
496        >Sith[i,Object(i),d(Object(i)),
497        >c(d(Object(i)))]=Sith[Object,d(Object),c(d(Object))]
498        >Sith[i,Object,d(Object(i))]=Sith[Object,d(Object)]
499        >Sith[i,Object(i)]=Sith[Object]
500            >Sits[F(i)]
501             >structure
502                 >Sith[Guess,s]
503                     >Sith[s,Subject(s)]
504                         >Sith[s,Subject,d(Subject(s))]
505                             >Sith[s,Subject(s),
506                             >d(Subject(s)),c(d(Subject(s)))]
507                         >Sith[s,Subject,d(Subject(s))] end
508                     >Sith[s,Subject(s)] end
509                     >Sith[s,Object(s)]
510                         >Sith[s,Object,d(Object(s))]
511                             Sith[s,Object(s),d(Object(s)),c(d(Object(s)))]
512                         >Sith[s,Object,d(Object(s))] end
513                     >Sith[s,Object(s)] end
514                 >Sith[Guess,s] end
515             >structure end
516             >Konw[Causality]
517              >of Sith[s,Subject(s)], Sith[s,Subject(s),d(Subject(s))],
518              >Sith[Subject(s),d(Subject(s)),c(d(Subject(s)))]
519              >Sith[s,Object(s)], Sith[s,Object(s),d(Object(s))],
520              >Sith[s,Object(s),d(Object(s)),c(d(Object(s)))]
521             >Konw[Causality]end
522            >Sits[F(i)]end
523        >w1,w2=groups(m)
524         >w1\=w1(s)
525         >w2\=w2(s)
526        >Konr[w1,w2]
527          >Sith[C,w1]
528          >Sith[C,w2]
529         >Konr[w1,w2]end
530         >Sits[G]=Konf[int](Konr[w1,w2])
531         >E[Ч,e[Ч](Subject)]=Sith[C,w1]
532          >E[Ч,e[Ч](Object)]=Sith[C,w2]
533        >Konw[Causality]=Class(Konr[w1,w2])
534        >Sits[G]
535         >E[Ц,e[Ц](G)]\=E[Ц,e[Ц](F)]
536        >if (abs(Sits[G])\=0):
537         >Ex1loop=1
538        >elif (abs(Sits[G])=0):
539         >Ex1loop=0
540         >proceed
```

```
541        >endif
542           >Sits[G]
543            >structure
544               >Sith[Subject]
545                    >Sith[Subject,d(Subject)]
546                         >Sith[Subject,d(Subject),c(d(Subject))]
547                    >Sith[Subject,d(Subject)] end
548               >Sith[Subject] end
549               >Sith[Object]
550                    >Sith[Object,d(Object)]
551                         >Sith[Object,d(Object),c(d(Object))]
552                    >Sith[Object,d(Object)] end
553               >Sith[Object] end
554           >structure end
555           >Konw[Causality]
556            >of Sith[Subject], Sith[Subject,d(Subject)],
557            >Sith[Subject,d(Subject),c(d(Subject))]
558            >Sith[Object], Sith[Object,d(Object)], Sith[Object,d(Object),c(d(Object))]
559            >Konw[Causality]end
560          >Sits[G] end
561     >Sith[G]=Konw[M](Sits[G])
562      >E[Ч,e[Ч](G)]=E[3,e[3](G)]
563     >Kont[1(i)]
564      >i=i+1
565     >endif
566     >else
567      >\Sits[R]
568     >Sits[Rnew]\=Sits[R]
569          >Sits[Rnew]
570           >structure
571               >Sith[Rnew]; Fin[Rnew]
572                    >Sith[Rnew,anew]; Fin[Rnew,anew]
573                         >Sith[Rnew,anew,bnew(anew)]
574                    >Sith[Rnew,anew]; Fin[Rnew,anew] end
575               >Sith[Rnew]; Fin[Rnew] end
576           >strucutre end
577          >Sit[Rnew] end
578     >Sits[R]=Sits[Rnew]
579      >E[Ц,e[Ц](R)=E[Ц,e[Ц](Rnew)
580     >endloop
581 >KonF[Ex1]end
582 >Kont[1(i)]
583          >Sits[3(i)]
584           >structure
585               >Sits[R]
586                >E[Ц,e[Ц](R)]
587               >Sits[G]
588                >E[Ц,e[Ц](G)]
589               >Sith[Obs3(i)]; Fin[Obs]
590               >Sith[C]
591                    >Sith[C,w]
592               >Sith[C] end
593               >Sith[G]
594                    >Sith[Subject]
595                         >Sith[Subject,d(Subject)]
596                              >Sith[Subject,d(Subject),c(d(Subject))]
597                         >Sith[Subject,d(Subject)] end
598                    >Sith[Subject] end
599                    >Sith[Object]
600                         >Sith[Object,d(Object)]
```

```
601                                            >Sith[Object,d(Object),c(d(Object))]
602                                        >Sith[Object,d(Object)] end
603                                    >Sith[Object] end
604                        >Sits[G] end
605                    >structure end
606                    >Konw[Conc(C,G)]
607                     >E[Ч,e[Ч](C)]
608                     >E[Ч,e[Ч](G)]
609                    >Konw[Conc(C,G)]end
610                    >Konw[Causality]
611                     >of Sith[Subject], Sith[Subject,d(Subject)],
612                     >Sith[Subject,d(Subject),c(d(Subject))]
613                     >Sith[Object], Sith[Object,d(Object)], Sith[Object,d(Object),c(d(Object))]
614                    >Konw[Causality]end
615                    >Konp[Obs3(i)]
616                     >of Sith[Obs3(i)], Fin[Obs3(i)]
617                     >E[Ц,e[Ц](3)], E[Ц,e[Ц](R)]
618                    >Konp[Obs3(i)]end
619                    >Konr[Descript(G,R)]
620                     >E[Ч,e[Ч](G)]
621                     >E[Ц,e[Ц](R)]
622                    >Konr[Descript(G,R)]end
623                >Sits[3(i)]end
624        >Kont[1(i)]end
625        >Kont[2]
626            >Sits[4]
627                >structure
628                    >Sits[R2]
629                     >E[Ц,e[Ц](R2)]
630                    >Sith[Obs4]; Fin[Obs]
631                    >Sith[I]; Fin[M]
632                        >Sith[I,u(I)]; Fin[M,u(M)]
633                    >Sith[I]; Fin[M] end
634                >structure end
635                >Konp[Obs4]
636                 >of Sith[Obs4], Fin[Obs]
637                 >E[Ц,e[Ц](4)], E[Ц,e[Ц](R2)]
638                >Konp[Obs4]end
639                >Konr[Descript(R2,4)]
640                 >E[Ч,e[Ч](M)], E[Ч,e[Ч](I)]
641                 >E[Ц,e[Ц](R2)]
642                >Konr[Descript(R2,4)]end
643            >Sits[4]end
644        >Konw[4,5]
645         >of Sits[4]
646         >restructure
647         >Sits[5]
648        >Konw[4,5]end
649            >Sits[5]
650                >structure
651                    >Sits[R2]
652                     >E[Ц,e[Ц](R2)]
653                    >Sith[Obs5]; Fin[Obs]
654                    >Sith[T]; Fin[T]
655                        >Sith[I]; Fin[M]
656                            >Sith[I,u(I)]; Fin[M,u(M)]
657                        >Sith[I]; Fin[M] end
658                    >Sith[T]; Fin[T] end
659                >structure end
660                >Konp[Obs5]
```

```
661                  >of Sith[Obs5], Fin[Obs]
662                   >Е[Ц,е[Ц](5)], Е[Ц,е[Ц](R2)]
663                  >Konp[Obs5]end
664                  >Konr[Descript(T,R2)]
665                   >Е[Ч,е[Ч](T)]
666                   >Е[Ц,е[Ц](R2)]
667                  >Konr[Descript(T,R2)]end
668                >Sits[5]end
669        >Kont[2]end
670        >KonF[Ex2]
671         >of Kont[2]
672         >Sits[Ex2]
673          >Е[Ц,е[Ц](Ex2)]
674         >Konr[Descript(T,Ex2)]
675          >Е[Ч,е[Ч](T)]
676          >Е[Ц,е[Ц](Ex2)]
677         >Konr[Descript(T,Ex2)]end
678         >if (Konr[Descript(T,Ex2)]=omnivial):
679          >Sits[ExP]
680           >Sits[R2in]=Sits[R2]
681            >Е[Ц,е[Ц](R2in)]=Е[Ц,е[Ц](R2)]
682           >Sits[Ex2]
683            >Е[Ц,е[Ц](Ex2)]
684           >Е[Ц,е[Ц](ExP)]=Е[Ц,е[Ц](R2in)]+Е[Ц,е[Ц](Ex2)]
685          >Sits[ExP]end
686          >Sits[R2]=Sits[ExP]
687           >Е[Ц,е[Ц](R2)]=Е[Ц,е[Ц](ExP)]
688          >outcome=2.0
689          >Kont[T]
690          >j=1
691          >Ex2A=1
692          >while (Ex2A=1):
693           >Kont[2(j)]
694           >KonF[Ex2(j)]
695            >of Kont[2(j)]
696            >Sits[Ex2(j)]
697             >Е[Ц,е[Ц](Ex2(j))]
698            >Konr[Ex2(j)]
699             >Е[Ч,е[Ч](T)]
700             >Е[Ц,е[Ц](Ex2(j))]
701            >Konr[Ex2(j)]end
702            >if (Konr[Ex2(j)]=omnivial):
703             >Sits[ExP(j)]
704              >Sits[R2in]=Sits[R2]
705               >Е[Ц,е[Ц](R2in)]=Е[Ц,е[Ц](R2)]
706              >Sits[Ex2(j)]
707               >Е[Ц,е[Ц](Ex2(j))]
708              >Е[Ц,е[Ц](ExP(j))]=Е[Ц,е[Ц](R2in)]+Е[Ц,е[Ц](Ex2(j))]
709             >Sits[ExP(j)]end
710             >Sits[R2]=Sits[ExP(j)]
711              >Е[Ц,е[Ц](R2)]=Е[Ц,е[Ц](ExP(j))]
712             >Ex2A=1
713             >outcome=2.1
714             >Kont[T]
715            >else
716             >Ex2A=0
717            >endif
718           >KonF[Ex2(j)]end
719          >j=j+1
720          >else
```

```
721        >Sits[ExP]
722         >Sits[R2in]=Sits[R2]
723          >E[Ц,e[Ц](R2in)]=E[Ц,e[Ц](R2)]
724         >Sits[Ex2(j)]
725          >E[Ц,e[Ц](Ex2(j))]
726        >E[Ц,e[Ц](ExP)]=E[Ц,e[Ц](R2in)]+E[Ц,e[Ц](Ex2(j))]
727        >Sits[ExP]end
728        >Sits[R3]=Sits[ExP]
729         >E[Ц,e[Ц](R3)]=E[Ц,e[Ц](ExP)]
730        >p2(Io)=v(I)
731        >p2(Mo)=v(M)
732        >Sith[Io,x(Io),x2(Io)]=Sith[I,u(I)]
733        >Fin[Mo,x(Mo),x2(Mo)]=Fin[M,u(M)]
734        >Sith[Io,x(Io)]=Sith[I]
735         >x(Io)=1
736        >Fin[Mo,x(Mo)]=Fin[M]
737         >x(Mo)=1
738        >Kont[3]
739       >endloop
740      >else
741       >Sits[ExP]
742        >Sits[R2in]=Sits[R2]
743         >E[Ц,e[Ц](R2in)]=E[Ц,e[Ц](R2)]
744        >Sits[Ex2]
745         >E[Ц,e[Ц](Ex2)]
746        >E[Ц,e[Ц](ExP)]=E[Ц,e[Ц](R2in)]+E[Ц,e[Ц](Ex2)]
747       >Sits[ExP]end
748       >Sits[R3]=Sits[ExP]
749      >E[Ц,e[Ц](R3)]=E[Ц,e[Ц](ExP)]
750       >p2(Io)=v(I)
751       >p2(Mo)=v(M)
752       >Sith[Io,x(Io),x2(Io)]=Sith[I,u(I)]
753       >Fin[Mo,x(Mo),x2(Mo)]=Fin[M,u(M)]
754       >Sith[Io,x(Io)]=Sith[I]
755        >x(Io)=1
756       >Fin[Mo,x(Mo)]=Fin[M]
757        >x(Mo)=1
758       >Kont[3]
759      >endif
760     >KonF[Ex2]end
761     >Kont[2(j)]
762           >Sits[5(j)]
763            >structure
764                  >Sits[R2]
765                   >E[Ц,e[Ц](R2)]
766                  >Sith[Obs5]; Fin[Obs]
767                  >Sith[T]; Fin[T]
768                      >Sith[I]; Fin[M]
769                          >Sith[I,u(I)]; Fin[M,u(M)]
770                      >Sith[I]; Fin[M] end
771                  >Sith[T]; Fin[T] end
772          >structure end
773          >Konp[Obs5(j)]
774           >of Sith[Obs5(j)], Fin[Obs]
775           >E[Ц,e[Ц](5(j))], E[Ц,e[Ц](R2)]
776          >Konp[Obs5(j)]end
777          >Konr[Descript(T,R2)]
778           >E[Ч,e[Ч](T)]
779           >E[Ц,e[Ц](R2)]
780          >Konr[Descript(T,R2)]end
```

```
781              >Sits[5(j)]end
782      >Kont[2(j)]end
783      >Kont[3]
784              >Sits[6A-1]
785               >structure
786                    >Sits[R3]
787                     >E[Ц,e[Ц](R3)]
788                    >Sith[Obs6A]; Fin[Obs]
789                    >Sith[T]; Fin[T]
790                         >Sith[To]; Fin[To]
791                              >Sith[Io]; Fin[Mo]
792                                   >Sith[Io,x(Io)]; Fin[Mo,x(Mo)]
793                                        >Sith[Io,x(Io),x2(x(Io))];
794                                        >Fin[Mo,x(Mo),x2(x(Mo))]
795                                   >Sith[Io,x(Io)]; Fin[Mo,x(Mo)] end
796                              >Sith[Io]; Fin[Mo] end
797                         >Sith[To]; Fin[To] end
798                    >Sith[T]; Fin[T] end
799               >structure end
800               >Konp[Obs6A-1]
801                >of Sith[Obs6A], Fin[Obs]
802                >E[Ц,e[Ц](6A-1)], E[Ц,e[Ц](R3)]
803               >Konp[Obs6A-1]end
804               >Konr[Descript(T,R3)]
805                >E[Ч,e[Ч](T)]
806                >E[Ц,e[Ц](R3)]
807               >Konr[Descript(T,R3)]end
808               >Konr[Prior,Sits[R3]]
809                >Fin[Mo,x(Mo,p)]
810                >Fin[Mo,x(Mo,p)+1]
811               >Konr[Prior,Sits[R3]]end
812              >Sits[6A-1]end
813      >Konw[6A-1,6A-2]
814       >of Sits[6A-1]
815       >Sith[C1A,cA]=mutual(Sith[R3,aA,bA(aA)])
816       >while (cA=0):
817        >\Sits[R3]
818        >Sits[R3new]\=Sits[R3]
819              >Sits[R3new]
820               >structure
821                    >Sith[R3new]; Fin[R3new]
822                         >Sith[R3new,aAnew]; Fin[R3new,aAnew]
823                              >Sith[R3new,aAnew,bAnew(aAnew)]
824                         >Sith[R3new,aAnew]; Fin[R3new,aAnew] end
825                    >Sith[R3new]; Fin[R3new] end
826               >strucutre end
827              >Sit[Rnew] end
828        >Sits[R3]=Sits[R3new]
829        >E[Ц,e[Ц](R3)=E[Ц,e[Ц](R3new)
830       >else:
831              >Sits[C1A]
832               >structure
833                    >Sith[C1A]
834                         >Sith[C1A,cA]
835                    >Sith[C1A] end
836               >structure end
837              >Sits[C1A] end
838        >Sith[dA]=Classes(Sith[C1A])
839              >Sits[C2A]
840               >structure
```

```
841                        >Sith[dA]
842                                >Sith[dA,cA(dA)]
843                        >Sith[dA] end
844                >structure end
845               >Sits[C2A] end
846        >Sith[Ci], Sith[Ci,yc]=Konw[M](Sits[C2A])
847         >E[Ч,e[Ч](Ci)]=E[3,e[3](C2A)]
848        >Sits[6A-2]
849      >endloop
850     >Konw[6A-1,6A-2]end
851             >Sits[6A-2]
852              >structure
853                    >Sits[R3]
854                     >E[Ц,e[Ц](R3)]
855                    >Sith[Obs6A-2]; Fin[Obs]
856                    >Sith[T]; Fin[T]
857                        >Sith[To]; Fin[To]
858                          >Sith[Io]; Fin[Mo]
859                            >Sith[Io,x(Io)]; Fin[Mo,x(Mo)]
860                                >Sith[Io,x(Io),x2(x(Io))];
861                                >Fin[Mo,x(Mo),x2(x(Mo))]
862                            >Sith[Io,x(Io)]; Fin[Mo,x(Mo)] end
863                          >Sith[Io]; Fin[Mo] end
864                        >Sith[To]; Fin[To] end
865                    >Sith[T]; Fin[T] end
866                    >Sith[Ci]
867                          >Sith[Ci,yc]
868                    >Sith[Ci] end
869              >structure end
870              >Konp[Obs6A-2]
871               >of Sith[Obs6A-2]; Fin[Obs]
872               >E[Ц,e[Ц](6A-2)], E[Ц,e[Ц](R3)]
873              >Konp[Obs6A-2]end
874              >Konr[Descript(T,R3)]
875               >E[Ч,e[Ч](T)]
876               >E[Ц,e[Ц](R3)]
877              >Konr[Descript(T,R3)]end
878              >Konr[Prior,Sits[R3]]
879               >Sith[Ci,yc(p)]
880               >Sith[Ci,yc(p)+1]
881              >Konr[Prior,Sits[R3]]end
882              >Konr[Prior,Sits[R3]]
883               >Fin[Mo,x(Mo,p)]
884               >Fin[Mo,x(Mo,p)+1]
885              >Konr[Prior,Sits[R3]]end
886             >Sits[6A-2]end
887      >Konw[6A-2,3A]
888       >of Sits[6A-2]
889       >yc1,yc2=groups(qc)
890      >yc1\=yc2
891      >Konr[yc1,yc2]
892        >Sith[C,yc1]
893        >Sith[C,yc2]
894       >Konr[yc1,yc2]end
895       >Sits[GA]=Konf[int](Konr[yc1,yc2])
896       >E[Ч,e[Ч](Subject)]=Sith[C,yc1]
897        >E[Ч,e[Ч](Object)]=Sith[C,yc2]
898       >Konw[CausalityA]=Class(Konr[yc1,yc2])
899           >Sits[GA]
900             >structure
```

15

```
901                         >Sith[GA]
902                              >Sith[SubjectA]
903                                   >Sith[SubjectA,dA(SubjectA)]
904                                        >Sith[SubjectA,dA(SubjectA),cA(dA(SubjectA))]
905                                   >Sith[SubjectA,dA(Subject)] end
906                              >Sith[SubjectA] end
907                              >Sith[ObjectA]
908                                   >Sith[ObjectA,dA(ObjectA)]
909                                        >Sith[ObjectA,dA(ObjectA),cA(dA(ObjectA))]
910                                   >Sith[ObjectA,dA(ObjectA)] end
911                              >Sith[ObjectA] end
912                         >Sith[G3A] end
913                    >structure end
914                    >Konw[CausalityA]
915                     >of Sith[SubjectA], Sith[SubjectA,dA(SubjectA)],
916                     >Sith[SubjectA,dA(SubjectA),cA(dA(SubjectA))]
917                     >Sith[ObjectA], Sith[ObjectA,dA(ObjectA)],
918                     >Sith[ObjectA,dA(ObjectA),cA(dA(ObjectA))]
919                    >Konw[CausalityA]end
920                   >Sits[GA] end
921           >Sith[GA]=Konw[M](Sits[GA])
922            >E[Ч,e[Ч](GA)]=E[3,e[3](GA)]
923          >Sits[3A]
924         >Konw[6A-2,3A] end
925                 >Sits[3A]
926                   >structure
927                         >Sits[R3]
928                          >E[Ц,e[Ц](R3)]
929                         >Sith[Obs3A]; Fin[Obs]
930                         >Sith[T]; Fin[T]
931                              >Sith[To]; Fin[To]
932                                   >Sith[Io]; Fin[Mo]
933                                        >Sith[Io,x(Io)]; Fin[Mo,x(Mo)]
934                                             >Sith[Io,x(Io),x2(x(Io))];
935                                             >Fin[Mo,x(Mo),x2(x(Mo))]
936                                        >Sith[Io,x(Io)]; Fin[Mo,x(Mo)] end
937                                   >Sith[Io]; Fin[Mo] end
938                              >Sith[To]; Fin[To] end
939                         >Sith[T]; Fin[T] end
940                         >Sith[Ci]
941                              >Sith[Ci,yc]
942                         >Sith[Ci] end
943                         >Sith[GA]
944                              >Sith[SubjectA]
945                                   >Sith[SubjectA,dA(SubjectA)]
946                                        >Sith[SubjectA,dA(SubjectA),cA(dA(SubjectA))]
947                                   >Sith[SubjectA,dA(Subject)] end
948                              >Sith[SubjectA] end
949                              >Sith[ObjectA]
950                                   >Sith[ObjectA,dA(ObjectA)]
951                                        >Sith[ObjectA,dA(ObjectA),cA(dA(ObjectA))]
952                                   >Sith[ObjectA,dA(ObjectA)] end
953                              >Sith[ObjectA] end
954                         >Sith[GA] end
955                    >structure end
956                    >Konw[Conc(Ci,GA)]
957                     >E[Ч,e[Ч](Ci)]
958                     >E[Ч,e[Ч](GA)]
959                    >Konnw[Conc(C,GA)]end
960                    >Konw[CausalityA]
```

16

```
961              >of Sith[SubjectA], Sith[SubjectA,dA(SubjectA)],
962              >Sith[SubjectA,dA(SubjectA),cA(dA(SubjectA))]
963              >Sith[ObjectA], Sith[ObjectA,dA(ObjectA)],
964              >Sith[ObjectA,dA(ObjectA),cA(dA(ObjectA))]
965            >Konw[CausalityA]end
966            >Konp[Obs3A]
967             >of Sith[Obs3A], Fin[Obs]
968             >Е[Ц,е[Ц](3A)], Е[Ц,е[Ц](R3)]
969            >Konp[Obs3A]end
970            >Konr[Descript((T,GA),R3)]
971             >Е[Ч,е[Ч](T)], Е[Ч,е[Ч](GA)]
972             >Е[Ц,е[Ц](R3)]
973            >Konr[Descript((T,GA),R3)]end
974            >Konr[Prior,Sits[R3]]
975             >Fin[Mo,x(Mo,p)]
976             >Fin[Mo,x(Mo,p)+1]
977            >Konr[Prior,Sits[R3]]end
978            >Konr[Prior,Sits[R3]]
979             >Sith[Ci,yc(p)]
980             >Sith[Ci,yc(p)+1]
981            >Konr[Prior,Sits[R3]]end
982          >Sits[3A]end
983      >Kont[3]end
984      >KonF[Ex3]
985       >Sits[Ex3]=Sits[R3]
986        >Е[Ц,е[Ц](Ex3)]=Е[Ц,е[Ц](R3)]
987       >Konr[Descript((T,GA),Ex1)]
988        >Е[Ч,е[Ч](T)], Е[Ч,е[Ч](GA)]
989        >Е[Ц,е[Ц](Ex1)]
990       >Konr[Descript((T,GA),Ex1)]end
991       >if (Konr[Descript((T,GA),Ex1)]=omnivial):
992        >Ex3A=1
993       >else:
994        >Ex3A=0
995       >endif
996       >Ex3loop=1
997       >iA=1
998       >while (Ex3loop=1):
999        >if (Ex3A=1):
1000         >Sits[R4]=Sits[Ex3]
1001          >Е[Ц,е[Ц](R4)]=Е[Ц,е[Ц](Ex3)]
1002             >Sits[ModelA]
1003              >structure
1004                  >Sith[chi]; Fin[chi]
1005                      >Sith[SubjectA]
1006                          >Sith[SubjectA,dA(SubjectA)]
1007                              >Sith[SubjectA,dA(SubjectA),cA(dA(SubjectA))]
1008                          >Sith[SubjectA,dA(SubjectA)] end
1009                      >Sith[SubjectA] end
1010                      >Sith[ObjectA]
1011                          >Sith[ObjectA,dA(ObjectA)]
1012                              >Sith[ObjectA,dA(ObjectA),cA(dA(ObjectA))]
1013                          >Sith[ObjectA,dA(ObjectA)] end
1014                      >Sith[ObjectA] end
1015                  >Sith[chi]; Fin[chi] end
1016              >structure end
1017              >Konw[NGA]
1018               >of Sith[SubjectA], Sith[SubjectA,dA(SubjectA)],
1019               >Sith[SubjectA,dA(SubjectA),cA(dA(SubjectA))]
```

```
1020              >Sith[ObjectA], Sith[ObjectA,dA(ObjectA)],
1021                >Sith[ObjectA,dA(ObjectA),cA(dA(ObjectA))]
1022             >Konw[NGA] end
1023           >Sits[ModelA] end
1024           >Sits[h,Ii]
1025            >structure
1026                >Sith[chi]; Fin[chi]
1027                     >Sith[SubjectA]
1028                          >Sith[SubjectA,dA(SubjectA)]
1029                               >Sith[SubjectA,dA(SubjectA),cA(dA(SubjectA))]
1030                          >Sith[SubjectA,dA(SubjectA)] end
1031                     >Sith[SubjectA] end
1032                >Sith[chi]; Fin[chi] end
1033            >structure end
1034           >Sits[h,Ii] end
1035        >Class(Sith[chi],Fin[chi](Sits[h,I]))=Konf[M](E[Ц,e[Ц](ModelA)],Konw[NGA])
1036        >Konr[SubjectA,chi]
1037          >Sith[SubjectA]
1038          >Sith[chi]
1039         >Konr[SubjectA,chi]end
1040        >Konr[SubjectA,chi]=Konf[M](Konw[NG])
1041        >Class(Konr[SubjectA,chi])=Ef
1042        >Sith[Ii], Sith[Ii,y(Ii)], Sith[Ii,y(Ii),y2(Ii)]
1043         >E[Ч^3,e[Ч^3](I)]=Class(Sith[chi],Fin[chi])
1044        >Fin[Mi], Fin[Mi,y(Mi)]=Konw[M](Sits[ModelA])
1045         >E[Ч^3,e[Ч^3](Mi)]=E[3,e[3](ModelA)]
1046        >q2(I)=abs(Class(Sith[chi],Fin[chi]))
1047        >q2(M)=abs(ALL(E[3,e[3](ModelA)]))
1048        >Kont[4]
1049      >elif (Ex3A=0):
1050      >sA[1:iA]
1051      >yc1(iA)=yc1
1052       >yc2(iA)=yc2
1053      >Sith[iA,SubjectA(iA),dA(SubjectA(iA)),cA(dA(SubjectA(iA)))]=
1054      >Sith[SubjectA,dA(SubjectA),cA(dA(SubjectA))]
1055      >Sith[iA,SubjectA,dA(SubjectA(iA))]=Sith[SubjectA,dA(SubjectA)]
1056      >Sith[iA,SubjectA(iA)]=Sith[SubjectA]
1057      >Sith[iA,ObjectA(iA),dA(Object(iA)),cA(dA(ObjectA(iA)))]=
1058      >Sith[ObjectA,dA(ObjectA),cA(dA(ObjectA))]
1059      >Sith[iA,ObjectA,dA(Object(iA))]=Sith[ObjectA,dA(ObjectA)]
1060      >Sith[iA,ObjectA(iA)]=Sith[ObjectA]
1061          >Sits[F(iA)]
1062           >structure
1063                >Sith[GuessA,sA]
1064                     >Sith[sA,SubjectA(sA)]
1065                          >Sith[sA,SubjectA,dA(SubjectA(sA))]
1066                               >Sith[sA,SubjectA(sA),dA(SubjectA(sA)),
1067                               >cA(dA(SubjectA(sA)))]
1068                          >Sith[sA,SubjectA,dA(SubjectA(sA))] end
1069                     >Sith[sA,SubjectA(sA)] end
1070                     >Sith[sA,ObjectA(sA)]
1071                          >Sith[sA,ObjectA,dA(ObjectA(sA))]
1072                               >Sith[sA,ObjectA(sA),dA(ObjectA(sA)),
1073                               >cA(dA(ObjectA(sA)))]
1074                          >Sith[sA,ObjectA,dA(ObjectA(sA))] end
1075                     >Sith[sA,ObjectA(sA)] end
1076                >Sith[GuessA,sA] end
1077           >structure end
1078           >Konw[CausalityA]
```

```
1079            >of Sith[sA,SubjectA(sA)], Sith[sA,SubjectA(sA),dA(SubjectA(sA))],
1080            >Sith[SubjectA(sA),dA(SubjectA(sA)),cA(dA(SubjectA(sA)))]
1081            >Sith[sA,ObjectA(sA)], Sith[sA,ObjectA(sA),dA(ObjectA(sA))],
1082            >Sith[sA,ObjectA(sA),dA(ObjectA(sA)),cA(dA(ObjectA(sA)))]
1083          >Konw[CausalityA]end
1084        >Sits[F(iA)]end
1085    >yc1,yc2=groups(qc)
1086     >yc1\=yc1(iA)
1087     >yc2\=yc2(iA)
1088    >Konr[yc1,yc2]
1089      >Sith[C,yc1]
1090      >Sith[C,yc2]
1091    >Konr[yc1,yc2]end
1092    >Sits[G]=Konf[int](Konr[yc1,yc2])
1093    >E[Ч,e[Ч](SubjectA)]=Sith[C,yc1]
1094     >E[Ч,e[Ч](ObjectA)]=Sith[C,yc2]
1095    >Konw[CausalityA]=Class(Konr[yc1,yc2])
1096    >Sits[GA]
1097    >E[Ц,e[Ц](GA)]\=E[Ц,e[Ц](F(iA))]
1098    >if (abs(Sits[GA])\=0):
1099     >Ex3loop=1
1100    >elif (abs(Sits[GA])=0):
1101     >Ex3loop=0
1102     >proceed
1103    >endif
1104        >Sits[GA]
1105         >structure
1106            >Sith[SubjectA]
1107                >Sith[SubjectA,dA(SubjectA)]
1108                    >Sith[SubjectA,dA(SubjectA),cA(dA(SubjectA))]
1109                >Sith[SubjectA,dA(SubjectA)] end
1110            >Sith[SubjectA] end
1111            >Sith[ObjectA]
1112                >Sith[ObjectA,dA(ObjectA)]
1113                    >Sith[ObjectA,dA(ObjectA),cA(dA(ObjectA))]
1114                >Sith[ObjectA,dA(ObjectA)] end
1115            >Sith[ObjectA] end
1116         >structure end
1117         >Konw[CausalityA]
1118          >of Sith[SubjectA], Sith[SubjectA,dA(SubjectA)],
1119          >Sith[SubjectA,dA(SubjectA),cA(dA(SubjectA))]
1120          >Sith[ObjectA], Sith[ObjectA,dA(ObjectA)],
1121          >Sith[ObjectA,dA(ObjectA),cA(dA(ObjectA))]
1122         >Konw[CausalityA]end
1123        >Sits[GA] end
1124     >Sith[GA]=Konw[M](Sits[GA])
1125      >E[Ч,e[Ч](GA)]=E[3,e[3](GA)]
1126     >Kont[3(iA)]
1127     >iA=iA+1
1128   >endif
1129  >else
1130   >ExA(3,4)=0
1131   >KonF[Ex4]
1132  >endloop
1133 >KonF[Ex3]end
1134 >Kont[3(iA)]
1135        >Sits[3A(iA)]
1136         >structure
1137            >Sits[R3]
1138             >E[Ц,e[Ц](R3)]
```

19

```
1139                         >Sits[G]
1140                          >E[Ц,e[Ц](G)]
1141                         >Sith[Obs3A(iA)]; Fin[Obs]
1142                         >Sith[T]; Fin[T]
1143                             >Sith[To]; Fin[To]
1144                                 >Sith[Io]; Fin[Mo]
1145                                     >Sith[Io,x(Io)]; Fin[Mo,x(Mo)]
1146                                         >Sith[Io,x(Io),x2(x(Io))];
1147                                         >Fin[Mo,x(Mo),x2(x(Mo))]
1148                                     >Sith[Io,x(Io)]; Fin[Mo,x(Mo)] end
1149                                 >Sith[Io]; Fin[Mo] end
1150                             >Sith[To]; Fin[To] end
1151                         >Sith[T]; Fin[T] end
1152                         >Sith[Ci]
1153                             >Sith[Ci,yc]
1154                         >Sith[Ci] end
1155                         >Sith[GA]
1156                             >Sith[SubjectA]
1157                                 >Sith[SubjectA,dA(SubjectA)]
1158                                     >Sith[SubjectA,dA(SubjectA),cA(dA(SubjectA))]
1159                                 >Sith[SubjectA,dA(SubjectA)] end
1160                             >Sith[SubjectA] end
1161                             >Sith[ObjectA]
1162                                 >Sith[ObjectA,dA(ObjectA)]
1163                                     >Sith[ObjectA,dA(ObjectA),cA(dA(ObjectA))]
1164                                 >Sith[ObjectA,dA(ObjectA)] end
1165                             >Sith[ObjectA] end
1166                         >Sits[GA] end
1167                     >structure end
1168                     >Konw[Conc(Ci,GA)]
1169                      >E[Ч,e[Ч](Ci)]
1170                      >E[Ч,e[Ч](GA)]
1171                     >Konw[Conc(Ci,GA)]end
1172                     >Konw[CausalityA]
1173                      >of Sith[SubjectA], Sith[SubjectA,dA(SubjectA)],
1174                      >Sith[SubjectA,dA(SubjectA),cA(dA(SubjectA))]
1175                      >Sith[ObjectA], Sith[ObjectA,dA(ObjectA)],
1176                      >Sith[ObjectA,dA(ObjectA),cA(dA(ObjectA))]
1177                     >Konw[CausalityA]end
1178                     >Konp[Obs3A(iA)]
1179                      >of Sith[Obs3A(iA)], Fin[Obs(iA)]
1180                      >E[Ц,e[Ц](3A)], E[Ц,e[Ц](R3)]
1181                     >Konp[Obs3A(iA)]end
1182                     >Konr[Descript((T,GA),R3)]
1183                      >E[Ч,e[Ч](T)], E[Ч,e[Ч](GA)]
1184                      >E[Ц,e[Ц](R3)]
1185                     >Konr[Descript((T,GA),R3)]end
1186                     >Konr[Prior,Sits[R3]]
1187                      >Fin[Mo,x(Mo,p)]
1188                      >Fin[Mo,x(Mo,p)+1]
1189                     >Konr[Prior,Sits[R3]]end
1190                     >Konr[Prior,Sits[R3]]
1191                      >Sith[Ci,yc(p)]
1192                      >Sith[Ci,yc(p)+1]
1193                     >Konr[Prior,Sits[R3]]end
1194                 >Sits[3A(iA)]
1195         >Kont[3(iA)]end
1196         >Kont[4]
1197                 >Sits[4A]
1198                  >structure
```

```
1199                          >Sits[R4]
1200                           >E[Ц,e[Ц](R4)]
1201                          >Sith[Obs4A]; Fin[Obs]
1202                          >Sith[T], Fin[T]
1203                              >Sith[To]; Fin[To]
1204                                  >Sith[Io]; Fin[Mo]
1205                                      >Sith[Io,x(Io)]; Fin[Mo,x(Mo)]
1206                                          >Sith[Io,x(Io),x2(x(Io))];
1207                                          >Fin[Mo,x(Mo),x2(x(Mo))]
1208                                      >Sith[Io,x(Io)]; Fin[Mo,x(Mo)] end
1209                                  >Sith[Io]; Fin[Mo] end
1210                              >Sith[To]; Fin[To] end
1211                              >Sith[Ti]; Fin[Mi]
1212                                  >Sith[Ii]; Fin[Mi]
1213                                      >Sith[Ii,y(Ii)], Fin[Mi,y(Mi)]
1214                                          >Sith[Ii,y(Ii),y2(y(Ii))];
1215                                          >Fin[Mi,y(Mi),y2(y(Mi))]
1216                                      >Sith[Ii,y(Ii)], Fin[Mi,y(Mi)] end
1217                                  >Sith[Ii]; Fin[Mi] end
1218                              >Sith[Ti]; Fin[Mi] end
1219                          >Sith[T], Fin[T] end
1220                  >structure end
1221                  >Konp[Obs4A]
1222                   >of Sith[Obs4A], Fin[Obs]
1223                   >E[Ц,e[Ц](4A)], E[Ц,e[Ц](R4)]
1224                  >Konp[Obs4A]end
1225                  >Konr[Descript(T,R4)]
1226                   >E[Ч,e[Ч](T)]
1227                   >E[Ц,e[Ц](R4)]
1228                  >Konr[Descript(T,R4)]end
1229                  >Konr[Prior,Sits[R4]]
1230                   >Fin[Mo,x(Mo,p)]
1231                   >Fin[Mo,x(Mo,p)+1]
1232                  >Konr[Prior,Sits[R4]]end
1233                  >Konr[Prior,Sits[R4]]
1234                   >Fin[Mi,y(Mi,p)]
1235                   >Fin[Mi,y(Mi,p)+1]
1236                  >Konr[Prior,Sits[R4]]end
1237                 >Sits[4A]end
1238      >Konw[4A,5A]
1239       >of Sits[4A]
1240       >restructure
1241       >Sits[5A]
1242      >Konw[4A,5A]end
1243              >Sits[5A]
1244              >structure
1245                  >Sits[R4]
1246                   >E[Ц,e[Ц](R4)]
1247                  >Sith[Obs5A]; Fin[Obs]
1248                  >Sith[T]; Fin[T]
1249                      >Sith[To]; Fin[To]
1250                          >Sith[Io,x(Io)]; Fin[Mo,x(Mo)]
1251                              >Sith[Io,x(Io),x2(x(Io))];
1252                              >Fin[Mo,x(Mo),x2(x(Mo))]
1253                          >Sith[Io,x(Io)]; Fin[Mo,x(Mo)] end
1254                      >Sith[To]; Fin[To] end
1255                      >Sith[Ti]; Fin[Ti]
1256                          >Sith[Ii,y(Ii)], Fin[Mi,y(Mi)]
1257                              >Sith[Ii,y(Ii),y2(y(Ii))];
1258                              >Fin[Mi,y(Mi),y2(y(Mi))<]
```

```
1259                                >Sith[Ii,y(Ii)], Fin[Mi,y(Mi)] end
1260                        >Sith[Ti]; Fin[Ti] end
1261                    >Sith[T]; Fin[T] end
1262              >structure end
1263              >Konp[Obs5A]
1264               >of Sith[Obs5A], Fin[Obs]
1265               >E[Ц,e[Ц](5A)], E[Ц,e[Ц](R4)]
1266              >Konp[Obs5A]end
1267              >Konr[Descript(T,R4)]
1268               >E[Ч,e[Ч](T)]
1269               >E[Ц,e[Ц](R4)]
1270              >Konr[Descript(T,R4)]end
1271              >Konr[Prior,Sits[R4]]
1272               >Fin[Mo,x(Mo,p)]
1273               >Fin[Mo,x(Mo,p)+1]
1274              >Konr[Prior,Sits[R4]]end
1275              >Konr[Prior,Sits[R4]]
1276               >Fin[Mi,y(Mi,p)]
1277               >Fin[Mi,y(Mi,p)+1]
1278              >Konr[Prior,Sits[R4]]end
1279              >ExA(3,4)=1
1280             >Sits[5A]end
1281      >Kont[4]end
1282      >KonF[Ex4]
1283       >if (ExA(3,4)=1):
1284        >Sits[Ex4]
1285         >E[Ц,e[Ц](Ex4)]
1286        >Sits[Exp]
1287         >Sits[R4in]=Sits[R4]
1288          >E[Ц,e[Ц](R4in)]=E[Ц,e[Ц](R4)]
1289         >Sits[Ex4]
1290          >E[Ц,e[Ц](Ex4)]
1291         >E[Ц,e[Ц](Exp)]=E[Ц,e[Ц](R4in)]+E[Ц,e[Ц](Ex4)]
1292        >Sits[Exp]end
1293        >Sits[R4(l[0])]=Sits[Exp]
1294         >E[Ц,e[Ц](R4(l[0]))]=E[Ц,e[Ц](Exp)]
1295        >Sits[5A(l[0])]=Sits[5A]
1296         >E[Ц,e[Ц](5A(l[0]))]=E[Ц,e[Ц](5A)]
1297       >elif (ExA(3,4)=0):
1298        >\Sith[Ci], Fin[Ci]
1299        >\Sith[GA]
1300         >\Sits[F(iA)]
1301        >Sits[R4(l[0])]=Sits[Ex3]
1302        >Sits[5A(l[0])]=Sits[3A]
1303       >endif
1304       >Ex40=1
1305       >l[0]=1
1306       >while (Ex40=1):
1307        >Ex4A=1
1308        >l[1]=1
1309        >while (Ex4A=1):
1310         >outcome=4.0
1311         >Kont[T]
1312         >Konr[Descript(T,R4(l[0]))]
1313          >E[Ч,e[Ч](T)]
1314          >E[Ц,e[Ц](R4(l[0]))]
1315         >Konr[Descript(T,R4(l[0]))]end
1316         >if (Konr[Descript(T,R4(l[0]))]=omnivial):
1317          >Ex4A=1
1318         >else
```

22

```
1319          >Ex4A=0
1320         >endif
1321         >l[1]=l[1]+1
1322        >else
1323         >Ex4B=0
1324         >l[2]=1
1325         >while ((Ex4B=0) and (l[2]<=l[2,max])):
1326          >Sits[R5]=Sits[R4(l[0])]
1327           >E[Ч,e[Ч](R5)]=E[Ч,e[Ч](R4(l[0]))]
1328          >task=Belt
1329          >Konq[Rebuild]
1330          >Sits[R4(l[0])]=Sits[R5]
1331           >E[Ч,e[Ч](R4(l[0]))]=E[Ч,e[Ч](R5)]
1332          >Sits[5A(l[0])]=Sits[5A_BPrior]
1333           >E[Ч,e[Ч](5A(l[0]))]=E[Ч,e[Ч](5A_BPrior)]
1334          >Konr[Descript(T,R4(l[0]))]
1335           >E[Ч,e[Ч](T)]
1336           >E[Ц,e[Ц](R4(l[0]))]
1337          >Konr[Descript(T,R4(l[0]))]end
1338          >if (Konr[Descript(T,R4(l[0]))]=omnivial):
1339           >Ex4B=1
1340          >else
1341           >Ex4B=0
1342          >endif
1343          >if (q=0):
1344           >Ex4C=0
1345           >l[3]=1
1346           >while ((Ex4C=0) and (q=0)):
1347            >Sits[R5]=Sits[R4(l[0])]
1348             >E[Ч,e[Ч](R5)]=E[Ч,e[Ч](R4(l[0]))]
1349            >task=AdBelt
1350            >Konq[Rebuild]
1351            >Sits[R4(l[0])]=Sits[R5]
1352             >E[Ч,e[Ч](R4(l[0]))]=E[Ч,e[Ч](R5)]
1353            >Sits[5A(l[0])]=Sits[5A_AdB]
1354             >E[Ч,e[Ч](5A(l[0]))]=E[Ч,e[Ч](5A_AdB)]
1355            >Konr[Descript(T,R4(l[0]))]
1356             >E[Ч,e[Ч](T)]
1357             >E[Ц,e[Ц](R4(l[0]))]
1358            >Konr[Descript(T,R4(l[0]))]end
1359            >if (Konr[Descript(T,R4(l[0]))]=omnivial):
1360             >Ex4C=1
1361            >else
1362             >Ex4C=0
1363            >endif
1364            >l[3]=l[3]+1
1365           >else
1366            >Ex4B=Ex4C
1367            >proceed
1368           >endloop
1369          >elif (q>0):
1370           >proceed
1371          >endif
1372         >else
1373          >if (Ex4B=1):
1374           >Ex4O=Ex4B
1375          >elif (Ex4B=0):
1376           >Ex4D=0
1377           >l[4]=1
1378           >while ((Ex4D=0) and (l[4]<=l[4,max])):
```

23

```
1379        >Sits[R5]=Sits[R4(l[0])]
1380         >E[Ч,e[Ч](R5)]=E[Ч,e[Ч](R4(l[0]))]
1381        >task=Core
1382        >Konq[Rebuild]
1383        >Sits[R4(l[0])]=Sits[R5]
1384         >E[Ч,e[Ч](R4(l[0]))]=E[Ч,e[Ч](R5)]
1385        >Sits[5A(l[0])]=Sits[5A_CPrior]
1386         >E[Ч,e[Ч](5A(l[0]))]=E[Ч,e[Ч](5A_CPrior)]
1387        >Konr[Descript(T,R4(l[0]))]
1388         >E[Ч,e[Ч](T)]
1389         >E[Ц,e[Ц](R4(l[0]))]
1390        >Konr[Descript(T,R4(l[0]))]end
1391        >if (Konr[Descript(T,R4(l[0]))]=omnivial):
1392         >Ex4D=1
1393        >else
1394         >Ex4D=0
1395        >endif
1396        >if (p=0):
1397         >Ex4E=0
1398         >l[5]=0
1399         >while ((Ex4E=0) and (p=0)):
1400          >Sits[R5]=Sits[R4(l[0])]
1401           >E[Ч,e[Ч](R5)]=E[Ч,e[Ч](R4(l[0]))]
1402          >task=Core
1403          >Konq[Rebuild]
1404          >Sits[R4(l[0])]=Sits[R5]
1405           >E[Ч,e[Ч](R4(l[0]))]=E[Ч,e[Ч](R5)]
1406          >Sits[5A(l[0])]=Sits[5A_AdC]
1407           >E[Ч,e[Ч](5A(l[0]))]=E[Ч,e[Ч](5A_AdC)]
1408          >Konr[Descript(T,R4(l[0]))]
1409           >E[Ч,e[Ч](T)]
1410           >E[Ц,e[Ц](R4(l[0]))]
1411          >Konr[Descript(T,R4(l[0]))]end
1412          >if (Konr[Descript(T,R4(l[0]))]=omnivial):
1413           >Ex4E=1
1414          >else
1415           >Ex4E=0
1416          >endif
1417          >l[5]=l[5]+1
1418         >else
1419          >Ex4D=Ex4E
1420         >endloop
1421        >elif (p>0):
1422        >endif
1423        >l[4]=l[4]+1
1424       >else
1425        >Ex4O=Ex4D
1426       >endloop
1427      >endif
1428     >endloop
1429    >endloop
1430   >else
1431    >Sits[R]=Sits[R4(l[0])]
1432     >E[Ч,e[Ч](R)]=E[Ч,e[Ч](R4(l[0]))]
1433    >Kont[1]
1434   >endloop
1435  >KonF[Ex4]end
1436  >Kont[T]
1437   >if (outcome=2.0):
1438    >Sits[R]=Sits[R2]
```

```
1439         >E[Ц,e[Ц](R)]=E[Ц,e[Ц](R2)]
1440        >Sits[T]=Sits[5]
1441         >E[Ц,e[Ц](T)]=E[Ц,e[Ц](5)]
1442       >elif (outcome=2.1):
1443        >Sits[R]=Sits[R2]
1444         >E[Ц,e[Ц](R)]=E[Ц,e[Ц](R2)]
1445        >Sits[T]=Sits[5(j)]
1446         >E[Ц,e[Ц](T)]=E[Ц,e[Ц](5(j))]
1447       >elif (outcome=4.0):
1448        >Sits[R]=Sits[R4(1[0])]
1449         >E[Ц,e[Ц](R)]=E[Ц,e[Ц](R4(1[0]))]
1450        >Sits[T]=Sits[5A(1[0])]
1451         >E[Ц,e[Ц](T)]=E[Ц,e[Ц](5A(1[0]))]
1452       >endif
1453             >Sits[T]
1454              >strucutre
1455                   >Sits[R]
1456                    >E[Ц,e[Ц](R)]
1457                   >Sith[ObsT]; Fin[Obs]
1458                   >Sith[T]; Fin[T]
1459                       >Sith[To]; Fin[To]
1460                            >Sith[Io,x(Io)]; Fin[Mo,x(Mo)]
1461                                 >Sith[Io,x(Io),x2(x(Io))];
1462                                 >Fin[Mo,x(Mo),x2(x(Mo))]
1463                            >Sith[Io,x(Io)]; Fin[Mo,x(Mo)] end
1464                       >Sith[To]; Fin[To] end
1465                       >Sith[Ti]; Fin[Ti]
1466                            >Sith[Ii,y(Ii)], Fin[Mi,y(Mi)]
1467                                 >Sith[Ii,y(Ii),y2(y(Ii))];
1468                                 >Fin[Mi,y(Mi),y2(y(Mi))]
1469                            >Sith[Ii,y(Ii)], Fin[Mi,y(Mi)] end
1470                       >Sith[Ti]; Fin[Ti] end
1471                   >Sith[T]; Fin[T] end
1472              >structure end
1473             >Konp[ObsT]
1474              >of Sith[ObsT], Fin[Obs]
1475              >E[Ц,e[Ц](5A)], E[Ц,e[Ц](R4)]
1476             >Konp[ObsT]end
1477             >Konr[Descript(T,R)]
1478              >E[Ч,e[Ч](T)]
1479              >E[Ц,e[Ц](R)]
1480             >Konr[Descript(T,R)]end
1481             >Konr[Prior,Sits[R]]
1482              >Fin[Mo,x(Mo,p)]
1483              >Fin[Mo,x(Mo,p)+1]
1484             >Konr[Prior,Sits[R]]end
1485             >Konr[Prior,Sits[R]]
1486              >Fin[Mi,y(Mi,p)]
1487              >Fin[Mi,y(Mi,p)+1]
1488             >Konr[Prior,Sits[R]]end
1489            >Sits[T]end
1490      >Konf[T]
1491       >of Sits[T]
1492      >Sith[I,t]=Sith[T]
1493       >E[Ч,e[Ч](I,t)]=E[Ч,e[Ч](T)]
1494      >Sits[R,t]=Sits[R]
1495       >E[Ц,e[Ц](R,t)]=E[Ц,e[Ц](R)]
1496      >Sits[T,t]=Sits[T]
1497       >E[Ц,e[Ц](T,t)]=E[Ц,e[Ц](T)]
1498      >Konf[T]end
```

```
1499          >Kont[T]end
1500      >Konq[T,t]end}
1501    {>Konq[Crosstest(tau1,tau2)]
1502      >recall ALL (tau1,tau2(theta))
1503      >recall Sith[T,tau1], Sith[T,tau1]
1504       >Sith[I(tau1,tau2)]
1505        >E[Ч,e[Ч](I,(tau1,tau2))]=E[Ч,e[Ч](I,tau1)] + E[Ч,e[Ч](I,tau2)]
1506        >E[Ч,e[Ч](I,(tau1,tau2))]=E[Ч,e[Ч](I,(tau2,tau1))]
1507       >Sith[I(tau1,tau2)]=norm(Sith[I(tau1,tau2)])
1508       >Sith[I(tau1ortau2)]
1509        >E[Ч,e[Ч](I,(tau1ortau2))]=E[Ч,e[Ч](I,tau1)] or E[Ч,e[Ч](I,tau2)]
1510        >E[Ч,e[Ч](I,(tau1ortau2))]=E[Ч,e[Ч](I,(tau2ortau1))]
1511       >Sith[I(tau1andtau2)]
1512        >E[Ч,e[Ч](I,(tau1andtau2))]=E[Ч,e[Ч](I,tau1)] and E[Ч,e[Ч](I,tau2)]
1513        >E[Ч,e[Ч](I,(tau1andtau2))]=E[Ч,e[Ч](I,(tau2andtau1))]
1514       >Fin[M(tau1ortau2)]
1515        >E[Ч,e[Ч](M,(tau1ortau2))]=E[Ч,e[Ч](M,tau1)] or E[Ч,e[Ч](M,tau2)]
1516        >E[Ч,e[Ч](M,(tau1ortau2))]=E[Ч,e[Ч](M,(tau2ortau1))]
1517       >Fin[M(tau1andtau2)]
1518        >E[Ч,e[Ч](M,(tau1andtau2))]=E[Ч,e[Ч](M,tau1)] and E[Ч,e[Ч](M,tau2)]
1519        >E[Ч,e[Ч](M,(tau1andtau2))]=E[Ч,e[Ч](M,(tau2andtau1))]
1520       >if (Konw[M-](Sith[I(tau1,tau2)])=Konw[M-](Sith[I(tau1)]):
1521       >Fin[M(tau1,tau2)]=Fin[M(tau1)]
1522       >Ct(tau1,tau2)=1
1523      >elif ((Konw[M-](Sith[I(tau1,tau2)])=Konw[M-](Sith[I(tau2)]):
1524       >Fin[M(tau1,tau2)]=Fin[M(tau2)]
1525       >Ct(tau1,tau2)=1
1526      >elif ((Konw[M-](Sith[I(tau1,tau2)])=Konw[M-](Sith[I(tau1andtau2)]])):
1527       >Fin[M(tau1,tau2)]=Fin[M(tau1andtau2)]
1528       >Ct(tau1,tau2)=1
1529      >elif ((Konw[M-](Sith[I(tau1,tau2)])=Konw[M-](Sith[I(tau1ortau2)]])):
1530       >Fin[M(tau1,tau2)]=Fin[M(tau1ortau2)]
1531       >Ct(tau1,tau2)=1
1532      >else
1533      >Kont[RRM]
1534       >Sits[R]
1535        >Sits[R,tau1]
1536         >E[Ц,e[Ц](R,tau1)]
1537        >Sits[R,tau2]
1538         >E[Ц,e[Ц](R,tau2)]
1539        >E[Ц,e[Ц](R)]=E[Ц,e[Ц](R,tau1)]+E[Ц,e[Ц](R,tau2)]
1540       >Sits[h,I]=Konw[M](Sith[I(tau1,tau2)])
1541        >E[Ц,e[Ц](h,I)]
1542                >Sits[RRM1]
1543                 >structure
1544                  >Sits[R]
1545                   >stucture
1546                      >Sith[R]; Fin[R]
1547                             >Sith[R,a]; Fin[R,a]
1548                                    >Sith[R,a,b(a)]
1549                             >Sith[R,a]; Fin[R,a] end
1550                      >Sith[R]; Fin[R] end
1551                 >stucture end
1552                >Sits[R] end
1553                >Sits[h,I]
1554                 >E[Ц,e[Ц](h,I)]
1555                     >Sith[Obs1]; Fin[Obs]
1556             >structure end
1557             >Konp[Obs1]
1558              >of Sith[Obs1], Fin[Obs]
```

```
1559              >E[Ц,e[Ц](RRM1)], E[Ц,e[Ц](R)], E[Ц,e[Ц](h,I)]
1560             >Konp[Obs1]end
1561            >Sits[RRM1]end
1562    >Konw[RRM1,RMM2]
1563     >of Sits[RRM1]
1564     >Siht[C1,c]= mutual Sith[R,a,b(a)]\ E[Ч,e[Ч](Subject)](Sits[h,I])
1565     >if (c=0):
1566      >proceed Konq[Crosstest(tau1,tau2)]
1567     >else
1568            >Sits[C1]
1569             >structure
1570                 >Sith[C1]
1571                      >Sith[C1,c]
1572                 >Sith[C1] end
1573             >structure end
1574            >Sits[C1] end
1575     >Sith[d]=Classes(Sith[C1])
1576            >Sits[C2]
1577             >structure
1578                 >Sith[d]
1579                      >Sith[d,c(d)]
1580                 >Sith[d] end
1581             >structure end
1582            >Sits[C2] end
1583     >Sith[C], Sith[C,w]=Konw[M](Sits[C2])
1584     >E[Ч,e[Ч](C)]=E[3,e[3](C2)]
1585     >Sits[RRM2]
1586     >endif
1587    >Konw[RRM1,RRM2]end
1588            >Sits[RRM2]
1589             >structure
1590              >Sits[R]
1591               >E[Ц,e[Ц](R)]
1592              >Sits[h,I]
1593               >E[Ц,e[Ц](h,I)]
1594                   >Sith[Obs2]; Fin[Obs]
1595                   >Sith[C]
1596                        >Sith[C,w]
1597                   >Sith[C] end
1598             >structure end
1599             >Konp[Obs2]
1600              >of Sith[Obs2], Fin[Obs]
1601              >E[Ц,e[Ц](RRM2)], E[Ц,e[Ц](R)], E[Ц,e[Ц](h,I)]
1602             >Konp[Obs2]end
1603            >Sits[RRM2]end
1604    >Konw[RRM2,RRM3]
1605     >of Sits[RRM2]
1606     >w1,w2=groups(m)
1607     >w1\=w2
1608     >Konr[w1,w2]
1609      >Sith[C,w1]
1610      >Sith[C,w2]
1611     >Konr[w1,w2]end
1612     >Sits[G]=Konf[int](Konr[w1,w2])
1613      >E[Ч,e[Ч](Subject)]=Sith[C,w1]
1614      >E[Ч,e[Ч](Object)]=Sith[C,w2]
1615     >Konw[Causality]=Class(Konr[w1,w2])
1616            >Sits[G]
1617             >structure
1618                 >Sith[Subject]
```

```
1619                                    >Sith[Subject,d(Subject)]
1620                                        >Sith[Subject,d(Subject),c(d(Subject))]
1621                                    >Sith[Subject,d(Subject)] end
1622                        >Sith[Subject] end
1623                        >Sith[Object]
1624                                    >Sith[Object,d(Object)]
1625                                        >Sith[Object,d(Object),c(d(Object))]
1626                                    >Sith[Object,d(Object)] end
1627                        >Sith[Object] end
1628                >structure end
1629                >Konw[Causality]
1630                 >of Sith[Subject], Sith[Subject,d(Subject)],
1631                 >Sith[Subject,d(Subject),c(d(Subject))]
1632                 >Sith[Object], Sith[Object,d(Object)],
1633                 >Sith[Object,d(Object),c(d(Object))]
1634                >Konw[Causality]end
1635            >Sits[G]end
1636      >Sith[G]=Konw[M](Sits[G])
1637       >E[Ч,e[Ч](G)]=E[3,e[3](G)]
1638      >Sits[RRM3]
1639    >Konw[RRM2,RRM3]end
1640                >Sits[RRM3]
1641                >structure
1642                >Sits[R]
1643                 >E[Ц,e[Ц](R)]
1644                >Sits[h,I]
1645                 >E[Ц,e[Ц](h,I)]
1646                    >Sith[Obs3]; Fin[Obs]
1647                    >Sith[C]
1648                            >Sith[C,w]
1649                    >Sith[C] end
1650                    >Sith[G]
1651                        >Sith[Subject]
1652                            >Sith[Subject,d(Subject)]
1653                                >Sith[Subject,d(Subject),c(d(Subject))]
1654                            >Sith[Subject,d(Subject)] end
1655                        >Sith[Subject] end
1656                        >Sith[Object]
1657                            >Sith[Object,d(Object)]
1658                                >Sith[Object,d(Object),c(d(Object))]
1659                            >Sith[Object,d(Object)] end
1660                        >Sith[Object] end
1661                    >Sith[G] end
1662                >structure end
1663                >Konw[Conc(C,G)]
1664                 >E[Ч,e[Ч](C)]
1665                 >E[Ч,e[Ч](G)]
1666                >Konw[Conc(C,G)]end
1667                >Konw[Causality]
1668                 >of Sith[Subject], Sith[Subject,d(Subject)],
1669                 >Sith[Subject,d(Subject),c(d(Subject))]
1670                 >Sith[Object], Sith[Object,d(Object)], Sith[Object,d(Object),c(d(Object))]
1671                >Konw[Causality]end
1672                >Konp[Obs2]
1673                 >of Sith[Obs2], Fin[Obs]
1674                 >E[Ц,e[Ц](RRM2)], E[Ц,e[Ц](R)], E[Ц,e[Ц](h,I)]
1675                >Konp[Obs2]end
1676                >Konr[Descript(G,R)]
1677                 >E[Ч,e[Ч](G)]
1678                 >E[Ц,e[Ц](R)]
```

28

```
1679            >Konr[Descript(G,R)]end
1680          >Sits[RRM3]end
1681      >Konf[ExRRM]
1682       >of Sits[RRM3]
1683       >Sits[ExRRM]=Sits[R]
1684        >E[Ц,е[Ц](ExRRM)]=E[Ц,е[Ц](R)]
1685       >Konr[ExRRM]
1686        >E[Ч,е[Ч](G)]
1687        >E[Ч,е[Ч](ExRRM)]
1688       >Konr[ExRRM]end
1689       >if (Konr[ExRRM]=omnivial):
1690        >ExRRMA=1
1691       >else
1692        >ExRRMA=0
1693       >endif
1694       >ExRRMloop=1
1695       >i_RRM=1
1696       >while (ExRRMloop=1), (i_RRM<=i_RRMmax):
1697        >if (ExRRMA=1):
1698              >Sits[Model]
1699               >structure
1700                    >Sith[chi]; Fin[chi]
1701                        >Sith[Subject]
1702                            >Sith[Subject,d(Subject)]
1703                                >Sith[Subject,d(Subject),c(d(Subject))]
1704                            >Sith[Subject,d(Subject)] end
1705                        >Sith[Subject] end
1706                        >Sith[Object]
1707                            >Sith[Object,d(Object)]
1708                                >Sith[Object,d(Object),c(d(Object))]
1709                            >Sith[Object,d(Object)] end
1710                        >Sith[Object] end
1711                    >Sith[chi]; Fin[chi] end
1712               >structure end
1713               >Konw[NG]
1714                >of Sith[Subject], Sith[Subject,d(Subject)],
1715                >Sith[Subject,d(Subject),c(d(Subject))]
1716                >Sith[Object], Sith[Object,d(Object)], Sith[Object,d(Object),c(d(Object))]
1717               >Konw[NG] end
1718              >Sits[Model] end
1719        >Class(Sith[chi],Fin[chi](Sits[h,I]))=Konf[M](E[Ц,е[Ц](Model)],Konw[NG])
1720        >Konr[Subject,chi]=Konf[M,R](Konw[NG])
1721        >Class(Konr[Subject,chi])=Ef
1722        >Sith[I], Sith[I,u(I)]
1723         >E[Ч,е[Ч](I)]=Class(Sith[chi],Fin[chi])
1724        >Fin[M], Fin[M,u(M)]=Konw[M](Sits[Model])
1725         >E[Ч,е[Ч](M)]=E[3,e[3](Model)]
1726        >v(I)=abs(Class(Sith[chi],Fin[chi]))
1727        >v(M)=abs(ALL(E[3,e[3](Model)]))
1728        >Ct(tau1,tau2)=1
1729        >Sits[RRM4]
1730        >elif (ExRRMA=0):
1731         >s[1:i_RRM]
1732        >w1(i)=w1
1733        >w2(i)=w2
1734         >Sith[i_RRM,Subject(i_RRM),d(Subject(i_RRM)),
1735         >c(d(Subject(i_RRM)))]=Sith[Subject,d(Subject),c(d(Subject))]
1736         >Sith[i_RRM,Subject,d(Subject(i_RRM))]=Sith[Subject,d(Subject)]
1737         >Sith[i_RRM,Subject(i_RRM)]=Sith[Subject]
```

29

```
1738        >Sith[i_RRM,Object(i_RRM),d(Object(i_RRM)),
1739        >c(d(Object(i_RRM)))]=Sith[Object,d(Object),c(d(Object))]
1740        >Sith[i_RRM,Object,d(Object(i_RRM))]=Sith[Object,d(Object)]
1741        >Sith[i_RRM,Object(i_RRM)]=Sith[Object]
1742            >Sits[F(i_RRM)]
1743             >structure
1744                >Sith[Guess,s]
1745                    >Sith[s,Subject(s)]
1746                        >Sith[s,Subject,d(Subject(s))]
1747                            >Sith[s,Subject(s),d(Subject(s)),
1748                            >c(d(Subject(s)))]
1749                        >Sith[s,Subject,d(Subject(s))] end
1750                    >Sith[s,Subject(s)] end
1751                    >Sith[s,Object(s)]
1752                        >Sith[s,Object,d(Object(s))]
1753                            >Sith[s,Object(s),d(Object(s)),
1754                            >c(d(Object(s)))]
1755                        >Sith[s,Object,d(Object(s))] end
1756                    >Sith[s,Object(s)] end
1757                >Sith[Guess,s] end
1758             >structure end
1759             >Konw[Causality]
1760              >of Sith[s,Subject(s)], Sith[s,Subject(s),
1761              >d(Subject(s))], Sith[Subject(s),d(Subject(s)),c(d(Subject(s)))]
1762              >Sith[s,Object(s)], Sith[s,Object(s),d(Object(s))],
1763              >Sith[s,Object(s),d(Object(s)),c(d(Object(s)))]
1764             >Konw[Causality]end
1765            >Sits[F(i_RRM)]end
1766    >w1,w2=groups(m)
1767    >w1\=w1(s)
1768    >w2\=w2(s)
1769    >Konr[w1,w2]
1770     >Sith[C,w1]
1771     >Sith[C,w2]
1772    >Konr[w1,w2]end
1773    >Sits[G]=Konf[int](Konr[w1,w2])
1774     >E[Ч,e[Ч](Subject)]=Sith[C,w1]
1775     >E[Ч,e[Ч](Object)]=Sith[C,w2]
1776    >Konw[Causality]=Class(Konr[w1,w2])
1777    >Sits[G]
1778    >E[Ц,e[Ц](G)]\=E[Ц,e[Ц](F)]
1779    >if (abs(Sits[G])\=0):
1780    >ExRRMloop=1
1781    >elif (abs(Sits[G])=0):
1782    >ExRRMloop=0
1783    >proceed
1784    >endif
1785            >Sits[G]
1786             >structure
1787                >Sith[Subject]
1788                    >Sith[Subject,d(Subject)]
1789                        >Sith[Subject,d(Subject),c(d(Subject))]
1790                    >Sith[Subject,d(Subject)] end
1791                >Sith[Subject] end
1792                >Sith[Object]
1793                    >Sith[Object,d(Object)]
1794                        >Sith[Object,d(Object),c(d(Object))]
1795                    >Sith[Object,d(Object)] end
1796                >Sith[Object] end
1797             >structure end
```

```
1798                    >Konw[Causality]
1799                     >of Sith[Subject], Sith[Subject,d(Subject)],
1800                     >Sith[Subject,d(Subject),c(d(Subject))]
1801                     >Sith[Object], Sith[Object,d(Object)],
1802                     >Sith[Object,d(Object),c(d(Object))]
1803                    >Konw[Causality]end
1804                  >Sits[G] end
1805           >Sith[G]=Konw[M](Sits[G])
1806            >E[Ч,e[Ч](G)]=E[3,e[3](G)]
1807           >Sits[3(i_RRM)]
1808           >i_RRM=i_RRM+1
1809          >endif
1810        >else
1811         >Ct(tau1,tau2)=0
1812         >proceed
1813        >endloop
1814      >Konf[ExRRM]end
1815                  >Sits[3(i_RRM)]
1816                    >structure
1817                        >Sits[R]
1818                         >E[Ц,e[Ц](R)]
1819                        >Sits[G]
1820                         >E[Ц,e[Ц](G)]
1821                        >Sith[Obs3(i)]; Fin[Obs]
1822                        >Sith[C]
1823                              >Sith[C,w]
1824                        >Sith[C] end
1825                        >Sith[G]
1826                            >Sith[Subject]
1827                                  >Sith[Subject,d(Subject)]
1828                                      >Sith[Subject,d(Subject),c(d(Subject))]
1829                                  >Sith[Subject,d(Subject)] end
1830                            >Sith[Subject] end
1831                            >Sith[Object]
1832                                  >Sith[Object,d(Object)]
1833                                      >Sith[Object,d(Object),c(d(Object))]
1834                                  >Sith[Object,d(Object)] end
1835                            >Sith[Object] end
1836                        >Sits[G] end
1837                    >structure end
1838                  >Konw[Conc(C,G)]
1839                   >E[Ч,e[Ч](C)]
1840                   >E[Ч,e[Ч](G)]
1841                  >Konw[Conc(C,G)]end
1842                  >Konw[Causality]
1843                   >of Sith[Subject], Sith[Subject,d(Subject)],
1844                   >Sith[Subject,d(Subject),c(d(Subject))]
1845                   >Sith[Object], Sith[Object,d(Object)], Sith[Object,d(Object),c(d(Object))]
1846                  >Konw[Causality]end
1847                  >Konp[Obs3(i_RRM)]
1848                   >of Sith[Obs3(i_RRM)], Fin[Obs3(i_RRM)]
1849                   >E[Ц,e[Ц](3)], E[Ц,e[Ц](R)]
1850                  >Konp[Obs3(i_RRM)]end
1851                  >Konr[Descript(G,R)]
1852                   >E[Ч,e[Ч](G)]
1853                   >E[Ц,e[Ц](G)]
1854                  >Konr[Descript(G,R)]end
1855                >Sits[3(i_RRM)]end
1856      >Kont[RRM]end
1857    >endif
```

```
1858    >Konq[Crosstest(tau1,tau2)]end}
1859  {>Konq[Ex4]
1860        >Kont[Ex4]
1861        >KonF[Ex4]
1862       >Sits[Ex4]
1863        >E[Ц,e[Ц](Ex4)]
1864       >Sits[Exp]
1865       >Sits[Rin]=Sits[R,t]
1866        >E[Ц,e[Ц](Rin)]=E[Ц,e[Ц](R,t)]
1867       >Sits[Ex4]
1868        >E[Ц,e[Ц](Ex4)]
1869       >E[Ц,e[Ц](Exp)]=E[Ц,e[Ц](Rin)]+E[Ц,e[Ц](Ex4)]
1870       >Sits[Exp]end
1871       >Sits[R4(l[0])]=Sits[Exp]
1872        >E[Ц,e[Ц](R4(l[0]))]=E[Ц,e[Ц](Exp)]
1873       >Sits[5A(l[0])]=Sits[T,t]
1874        >E[Ц,e[Ц](5A(l[0]))]=E[Ц,e[Ц](T,t)]
1875       >Ex4O=1
1876       >l[0]=1
1877       >while (Ex4O=1):
1878        >Ex4A=1
1879        >l[1]=1
1880        >while (Ex4A=1):
1881         >Sits[R,1]=Sits[R4(l[0])]
1882          >E[Ц,e[Ц](R,1)]=E[Ц,e[Ц](R4(l[0]))]
1883         >Sits[T,1]=Sits[5A(l[0])]
1884          >E[Ц,e[Ц](T,1)]=E[Ц,e[Ц](5A(l[0]))]
1885         >Konr[Descript(T,R4(l[0]))]
1886          >E[Ч,e[Ч](T)]
1887          >E[Ц,e[Ц](R4(l[0]))]
1888         >Konr[Descript(T,R4(l[0]))]end
1889         >if (Konr[Descript(T,R4(l[0]))]=omnivial):
1890          >Ex4A=1
1891         >else
1892          >Ex4A=0
1893         >endif
1894         >l[1]=l[1]+1
1895        >else
1896         >Ex4B=0
1897         >l[2]=1
1898         >while ((Ex4B=0) and (l[2]<=l[2,max])):
1899          >Sits[R5]=Sits[R4(l[0])]
1900           >E[Ч,e[Ч](R5)]=E[Ч,e[Ч](R4(l[0]))]
1901          >task=Belt
1902          >Konq[Rebuild]
1903          >Sits[R4(l[0])]=Sits[R5]
1904           >E[Ч,e[Ч](R4(l[0]))]=E[Ч,e[Ч](R5)]
1905          >Sits[5A(l[0])]=Sits[5A_BPrior]
1906           >E[Ч,e[Ч](5A(l[0]))]=E[Ч,e[Ч](5A_BPrior)]
1907          >Konr[Descript(T,R4(l[0]))]
1908           >E[Ч,e[Ч](T)]
1909           >E[Ц,e[Ц](R4(l[0]))]
1910          >Konr[Descript(T,R4(l[0]))]end
1911          >if (Konr[Descript(T,R4(l[0]))]=omnivial):
1912           >Ex4B=1
1913          >else
1914           >Ex4B=0
1915          >endif
1916          >if (q=0):
1917           >Ex4C=0
```

```
1918        >l[3]=1
1919        >while ((Ex4C=0) and (q=0)):
1920         >Sits[R5]=Sits[R4(l[0])]
1921          >E[Ч,e[Ч](R5)]=E[Ч,e[Ч](R4(l[0]))]
1922         >task=AdBelt
1923         >Konq[Rebuild]
1924         >Sits[R4(l[0])]=Sits[R5]
1925          >E[Ч,e[Ч](R4(l[0]))]=E[Ч,e[Ч](R5)]
1926         >Sits[5A(l[0])]=Sits[5A_AdB]
1927          >E[Ч,e[Ч](5A(l[0]))]=E[Ч,e[Ч](5A_AdB)]
1928         >Konr[Descript(T,R4(l[0]))]
1929          >E[Ч,e[Ч](T)]
1930          >E[Ц,e[Ц](R4(l[0]))]
1931         >Konr[Descript(T,R4(l[0]))]end
1932         >if (Konr[Descript(T,R4(l[0]))]=omnivial):
1933          >Ex4C=1
1934         >else
1935          >Ex4C=0
1936         >endif
1937         >l[3]=l[3]+1
1938        >else
1939         >Ex4B=Ex4C
1940         >proceed
1941        >endloop
1942       >elif (q>0):
1943        >proceed
1944       >endif
1945      >else
1946       >if (Ex4B=1):
1947        >Ex4O=Ex4B
1948       >elif (Ex4B=0):
1949        >Ex4D=0
1950        >l[4]=1
1951        >while ((Ex4D=0) and (l[4]<=l[4,max])):
1952         >Sits[R5]=Sits[R4(l[0])]
1953          >E[Ч,e[Ч](R5)]=E[Ч,e[Ч](R4(l[0]))]
1954         >task=Core
1955         >Konq[Rebuild]
1956         >Sits[R4(l[0])]=Sits[R5]
1957          >E[Ч,e[Ч](R4(l[0]))]=E[Ч,e[Ч](R5)]
1958         >Sits[5A(l[0])]=Sits[5A_CPrior]
1959          >E[Ч,e[Ч](5A(l[0]))]=E[Ч,e[Ч](5A_CPrior)]
1960         >Konr[Descript(T,R4(l[0]))]
1961          >E[Ч,e[Ч](T)]
1962          >E[Ц,e[Ц](R4(l[0]))]
1963         >Konr[Descript(T,R4(l[0]))]end
1964        >if (Konr[Descript(T,R4(l[0]))]=omnivial):
1965         >Ex4D=1
1966        >else
1967         >Ex4D=0
1968        >endif
1969        >if (p=0):
1970         >Ex4E=0
1971         >l[5]=0
1972         >while ((Ex4E=0) and (p=0)):
1973          >Sits[R5]=Sits[R4(l[0])]
1974           >E[Ч,e[Ч](R5)]=E[Ч,e[Ч](R4(l[0]))]
1975          >task=Core
1976          >Konq[Rebuild]
1977          >Sits[R4(l[0])]=Sits[R5]
```

33

```
1978          >E[Ч,e[Ч](R4(l[0]))]=E[Ч,e[Ч](R5)]
1979          >Sits[5A(l[0])]=Sits[5A_AdC]
1980           >E[Ч,e[Ч](5A(l[0]))]=E[Ч,e[Ч](5A_AdC)]
1981          >Konr[Descript(T,R4(l[0]))]
1982           >E[Ч,e[Ч](T)]
1983           >E[Ц,e[Ц](R4(l[0]))]
1984          >Konr[Descript(T,R4(l[0]))]end
1985          >if (Konr[Descript(T,R4(l[0]))]=omnivial):
1986           >Ex4E=1
1987          >else
1988           >Ex4E=0
1989          >endif
1990          >l[5]=l[5]+1
1991          >else
1992           >Ex4D=Ex4E
1993          >endloop
1994         >elif (p>0):
1995         >endif
1996         >l[4]=l[4]+1
1997        >else
1998         >Ex4O=Ex4D
1999        >endloop
2000       >endif
2001      >endloop
2002     >endloop
2003    >else
2004     >Sits[R,1]=Sits[R4(l[0])]
2005      >E[Ц,e[Ц](R,1)]=E[Ц,e[Ц](R4(l[0]))]
2006     >Sits[T,1]=Sits[5A(l[0])]
2007      >E[Ц,e[Ц](T,1)]=E[Ц,e[Ц](5A(l[0]))]
2008    >endloop
2009   >KonF[Ex4]end
2010   >Kont[Ex4]end
2011  >Konq[Ex4]end}
2012 {>Konq[Rebuild]
2013  >if (task=Belt):
2014   >if (abs(Sith[Ti],Fin[Ti])=1)):
2015    >parallel (Kont[AdBelt], Kont[ReBelt])
2016   >else
2017    >parallel (Kont[NiBelt], Kont[AdBelt], Kont[ReBelt])
2018   >endif
2019   >Kont[ordBelt]
2020  >elif (task=AdBelt):
2021   >Kont[AdBelt]
2022  >elif (task=Core):
2023   >if (abs(Sith[To],Fin[To])):
2024    >parallel (Kont[AdCore], Kont[ReCore]):
2025   >else
2026    >parallel (Kont[NiCore], Kont[AdCore], Kont[ReCore])
2027   >endif
2028   >Kont[ordCore]
2029  >elif (task=AdCore):
2030   >Kont[AdCore]
2031  >endif
2032      >Kont[NiBelt]
2033       >Sits[5A_NiB_1]=Sits[5A(l[0])]
2034       >E[Ц,e[Ц](5A_NiB_1)]=E[Ц,e[Ц](5A(l[0]))]
2035          >Sits[5A_NiB_1]
2036           >structure
2037               >Sits[R5]
```

```
2038                                      >E[Ц,e[Ц](R5)]
2039                                  >Sith[Obs5A_NiB_1]; Fin[Obs]
2040                                  >Sith[T]; Fin[T]
2041                                      >Sith[To]; Fin[To]
2042                                          >Sith[Io,x(Io)]; Fin[Mo,x(Mo)]
2043                                              >Sith[Io,x(Io),x2(x(Io))];
2044                                              >Fin[Mo,x(Mo),x2(x(Mo))]
2045                                          >Sith[Io,x(Io)]; Fin[Mo,x(Mo)] end
2046                                      >Sith[To]; Fin[To] end
2047                                      >Sith[Ti]; Fin[Ti]
2048                                          >Sith[Ii,y(Ii)], Fin[Mi,y(Mi)]
2049                                              >Sith[Ii,y(Ii),y2(y(Ii))];
2050                                              >Fin[Mi,y(Mi),y2(y(Mi))]
2051                                          >Sith[Ii,y(Ii)], Fin[Mi,y(Mi)] end
2052                                      >Sith[Ti]; Fin[Ti] end
2053                                  >Sith[T]; Fin[T] end
2054                              >structure end
2055                              >Konp[Obs5A_NiB_1]
2056                               >of Sith[Obs5A_NiB_1], Fin[Obs]
2057                               >E[Ц,e[Ц](5A_NiB_1)], E[Ц,e[Ц](R5)]
2058                              >Konp[Obs5A_NiB_1]end
2059                              >Konr[Descript(T,R5)]
2060                               >E[Ч,e[Ч](T)]
2061                               >E[Ц,e[Ц](R5)]
2062                              >Konr[Descript(T,R5)]end
2063                              >Konr[Prior,Sits[R5]]
2064                               >Sith[To]
2065                               >Sith[Ti]
2066                              >Konr[Prior,Sits[R5]]end
2067                              >Konr[Prior,Sits[R5]]
2068                               >Fin[Mo,x(Mo,p)]
2069                               >Fin[Mo,x(Mo,p)+1]
2070                              >Konr[Prior,Sits[R5]]end
2071                              >Konr[Prior,Sits[R5]]
2072                               >Fin[Mi,y(Mi,p)]
2073                               >Fin[Mi,y(Mi,p)+1]
2074                              >Konr[Prior,Sits[R5]]end
2075                          >Sits[5A_NiB_1]end
2076              >Konw[4NiBelt_2]
2077               >of Sits[5A_NiB_1]
2078               >Sits[5A_NiB_2]
2079               >y(Mi)[1:q(Mi)-1]
2080               >y(Mi,p)[1:q(Mi)-2]
2081               >Fin[Mi,y(Mi)]
2082                >\Fin[Mi,y(Mi,p)max+1]
2083               >q(Mi)=q(Mi)-1
2084               >y(Mi)[1:q(Mi)]
2085               >y(Mi,p)[1:q(Mi)-1]
2086              >Konw[4NiBelt_2]end
2087                      >Sits[5A_NiB_2]
2088                       >structure
2089                          >Sits[R5]
2090                           >E[Ц,e[Ц](R5)]
2091                          >Sith[Obs5A_NiB_2]; Fin[Obs]
2092                          >Sith[T]; Fin[T]
2093                              >Sith[To]; Fin[To]
2094                                  >Sith[Io,x(Io)]; Fin[Mo,x(Mo)]
2095                                      >Sith[Io,x(Io),x2(x(Io))];
2096                                      >Fin[Mo,x(Mo),x2(x(Mo))]
2097                                  >Sith[Io,x(Io)]; Fin[Mo,x(Mo)] end
```

```
2098                                      >Sith[To]; Fin[To] end
2099                                      >Sith[Ti]; Fin[Ti]
2100                                            >Sith[Ii,y(Ii)], Fin[Mi,y(Mi)]
2101                                                  >Sith[Ii,y(Ii),y2(y(Ii))];
2102                                                  >Fin[Mi,y(Mi),y2(y(Mi))]
2103                                            >Sith[Ii,y(Ii)], Fin[Mi,y(Mi)] end
2104                                      >Sith[Ti]; Fin[Ti] end
2105                              >Sith[T]; Fin[T] end
2106                   >structure end
2107                   >Konp[Obs5A_NiB_2]
2108                    >of Sith[Obs5A_NiB_2], Fin[Obs]
2109                    >E[Ц,e[Ц](5A_NiB_2)], E[Ц,e[Ц](R5)]
2110                   >Konp[Obs5A_NiB_2]end
2111                   >Konr[Descript(T,R5)]
2112                    >E[Ч,e[Ч](T)]
2113                    >E[Ц,e[Ц](R5)]
2114                   >Konr[Descript(T,R5)]end
2115                   >Konr[Prior,Sits[R5]]
2116                    >Sith[To]
2117                    >Sith[Ti]
2118                   >Konr[Prior,Sits[R5]]end
2119                   >Konr[Prior,Sits[R5]]
2120                    >Fin[Mo,x(Mo,p)]
2121                    >Fin[Mo,x(Mo,p)+1]
2122                   >Konr[Prior,Sits[R5]]end
2123                   >Konr[Prior,Sits[R5]]
2124                    >Fin[Mi,y(Mi,p)]
2125                    >Fin[Mi,y(Mi,p)+1]
2126                   >Konr[Prior,Sits[R5]]end
2127              >Sits[5A_NiB_2]end
2128         >Sits[5A_NiB]=Sits[5A_NiB_2]
2129          >E[Ц,e[Ц](5A_NiB)]=E[Ц,e[Ц](5A_NiB_2)]
2130      >Kont[NiBelt]end
2131      >Kont[AdBelt]
2132       >Sits[5A_AdB_1]=Sits[5A(1[0])]
2133        >E[Ц,e[Ц](5A_AdB_1)]=E[Ц,e[Ц](5A(1[0]))]
2134              >Sits[5A_AdB_1]
2135               >structure
2136                    >Sits[R5]
2137                     >E[Ц,e[Ц](R5)]
2138                    >Sith[Obs5A_AdB_1]; Fin[Obs]
2139                    >Sith[T]; Fin[T]
2140                          >Sith[To]; Fin[To]
2141                                >Sith[Io,x(Io)]; Fin[Mo,x(Mo)]
2142                                      >Sith[Io,x(Io),x2(x(Io))];
2143                                      >Fin[Mo,x(Mo),x2(x(Mo))]
2144                                >Sith[Io,x(Io)]; Fin[Mo,x(Mo)] end
2145                          >Sith[To]; Fin[To] end
2146                          >Sith[Ti]; Fin[Ti]
2147                                >Sith[Ii,y(Ii)], Fin[Mi,y(Mi)]
2148                                      >Sith[Ii,y(Ii),y2(y(Ii))];
2149                                      >Fin[Mi,y(Mi),y2(y(Mi))]
2150                                >Sith[Ii,y(Ii)], Fin[Mi,y(Mi)] end
2151                          >Sith[Ti]; Fin[Ti] end
2152                    >Sith[T]; Fin[T] end
2153               >structure end
2154               >Konp[Obs5A_AdB_1]
2155                >of Sith[Obs5A_AdB_1], Fin[Obs]
2156                >E[Ц,e[Ц](5A_AdB_1)], E[Ц,e[Ц](R5)]
2157               >Konp[Obs5A_AdB_1]end
```

```
2158          >Konr[Descript(T,R5)]
2159           >E[Ч,e[Ч](T)]
2160           >E[Ц,e[Ц](R5)]
2161          >Konr[Descript(T,R5)]end
2162          >Konr[Prior,Sits[R5]]
2163           >Sith[To]
2164           >Sith[Ti]
2165          >Konr[Prior,Sits[R5]]end
2166          >Konr[Prior,Sits[R5]]
2167           >Fin[Mo,x(Mo,p)]
2168           >Fin[Mo,x(Mo,p)+1]
2169          >Konr[Prior,Sits[R5]]end
2170          >Konr[Prior,Sits[R5]]
2171           >Fin[Mi,y(Mi,p)]
2172           >Fin[Mi,y(Mi,p)+1]
2173          >Konr[Prior,Sits[R5]]end
2174         >Sits[5A_AdB_1]end
2175     >Konw[4AdBelt_2]
2176      >of Sits[6A_AdB_1]
2177      >Sits[5A_AdB_2]
2178      >y(Mi)[1:q(Mi)-1]
2179      >y(Mi,p)[1:q(Mi)-2]
2180      >Fin[Mi]
2181      >\Fin[Mi,y(Mi,p)max+1]
2182      >Fin[Mi,AdBmax(Mi),y2(AdBmax(Mi))]=Fin[Mi,y(Mi,p)max+1,y2(y(Mi,p)max+1)]
2183      >Fin[Mi,AdBmax(Mi)]=Fin[Mi,y(Mi,p)max+1]
2184     >Konw[4AdBelt_2]end
2185          >Sits[5A_AdB_2]
2186           >structure
2187              >Sits[R5]
2188               >E[Ц,e[Ц](R5)]
2189              >Sith[Obs5A_AdB_2]; Fin[Obs]
2190              >Sith[T]; Fin[T]
2191                  >Sith[To]; Fin[To]
2192                      >Sith[Io,x(Io)]; Fin[Mo,x(Mo)]
2193                          >Sith[Io,x(Io),x2(x(Io))];
2194                          >Fin[Mo,x(Mo),x2(x(Mo))]
2195                      >Sith[Io,x(Io)]; Fin[Mo,x(Mo)] end
2196                  >Sith[To]; Fin[To] end
2197                  >Sith[Ti]; Fin[Ti]
2198                      >Sith[Ii1]; Fin[Mi1]
2199                          >Sith[Ii,y(Ii)], Fin[Mi,y(Mi)]
2200                              >Sith[Ii,y(Ii),y2(y(Ii))];
2201                              >Fin[Mi,y(Mi),y2(y(Mi))]
2202                          >Sith[Ii,y(Ii)], Fin[Mi,y(Mi)] end
2203                      >Sith[Ii1]; Fin[Mi1] end
2204                      >Sith[Ii,AdBmax(Ii)]; Fin[Mi,AdBmax(Mi)]
2205                          >Sith[Ii,AdBmax(Ii),y2(AdBmax(Ii))];
2206                          >Fin[Mi,y(Mi),y2(AdBmax(Mi))]
2207                      >Sith[Ii,AdBmax(Ii)]; Fin[Mi,AdBmax(Mi)] end
2208                  >Sith[Ti]; Fin[Ti] end
2209              >Sith[T]; Fin[T] end
2210           >structure end
2211          >Konp[Obs5A_AdB_2]
2212           >of Sith[Obs5A_AdB_2], Fin[Obs]
2213           >E[Ц,e[Ц](5A_AdB_2)], E[Ц,e[Ц](R5)]
2214          >Konp[Obs5A_AdB_2]end
2215          >Konr[Descript(T,R5)]
2216           >E[Ч,e[Ч](T)]
2217           >E[Ц,e[Ц](R5)]
```

```
2218                    >Konr[Descript(T,R5)]end
2219                    >Konr[Prior,Sits[R5]]
2220                     >Sith[To]
2221                     >Sith[Ti]
2222                    >Konr[Prior,Sits[R5]]end
2223                    >Konr[Prior,Sits[R5]]
2224                     >Fin[Mo,x(Mo,p)]
2225                     >Fin[Mo,x(Mo,p)+1]
2226                    >Konr[Prior,Sits[R5]]end
2227                    >Konr[Prior,Sits[R5]]
2228                     >Fin[Mi,y(Mi,p)]
2229                     >Fin[Mi,y(Mi,p)+1]
2230                    >Konr[Prior,Sits[R5]]end
2231                   >Sits[5A_AdB_2]end
2232            >Konw[4AdBelt_3]
2233             >of Sits[5A_AdB_2]
2234             >Sith[C1_AdB,c_AdB]=mutual(Sith[R5,a_AdB,b_AdB(a_AdB)])
2235             >while (c_AdB=0):
2236               >\Sits[R5]
2237              >Sits[R5new]\=Sits[R5]
2238                    >Sits[R5new]
2239                     >structure
2240                        >Sith[R5new]; Fin[R5new]
2241                            >Sith[R5new,a_AdBnew]; Fin[R5new,a_AdBnew]
2242                                >Sith[R5new,a_AdBnew,b_AdBnew(a_AdBnew)]
2243                            >Sith[R5new,a_AdBnew]; Fin[R5new,a_AdBnew] end
2244                        >Sith[R5new]; Fin[R5new] end
2245                     >strucutre end
2246                    >Sit[R5new] end
2247              >Sits[R5]=Sits[R5new]
2248               >Е[Ц,e[Ц](R5)=E[Ц,e[Ц](R5new)
2249             >else:
2250                    >Sits[C1_AdB]
2251                     >structure
2252                        >Sith[C1_AdB]
2253                            >Sith[C1_AdB,c_AdB]
2254                        >Sith[C1_AdB] end
2255                     >structure end
2256                    >Sits[C1_AdB] end
2257              >Sith[d_AdB]=Classes(Sith[C1_AdB])
2258                    >Sits[C2_AdB]
2259                     >structure
2260                        >Sith[d_AdB]
2261                            >Sith[d_AdB,c_AdB(d_AdB)]
2262                        >Sith[d_AdB] end
2263                     >structure end
2264                    >Sits[C2_AdB] end
2265              >Sith[C_AdB],Sith[C_AdB,w_AdB]=Konw[M](Sits[C2_AdB])
2266               >Е[Ч,e[Ч](C_AdB)]=E[3,e[3](C2_AdB)]
2267             >Sits[5A_AdB_3]
2268            >endloop
2269           >Konw[4AdBelt_3]end
2270                >Sits[5A_AdB_3]
2271                 >structure
2272                    >Sits[R5]
2273                     >Е[Ц,e[Ц](R5)]
2274                    >Sith[Obs5A_AdB_3]; Fin[Obs]
2275                    >Sith[T]; Fin[T]
2276                        >Sith[To]; Fin[To]
2277                            >Sith[Io,x(Io)]; Fin[Mo,x(Mo)]
```

```
2278                                                        >Sith[Io,x(Io),x2(x(Io))];
2279                                                       >Fin[Mo,x(Mo),x2(x(Mo))]
2280                                          >Sith[Io,x(Io)]; Fin[Mo,x(Mo)] end
2281                              >Sith[To]; Fin[To] end
2282                              >Sith[Ti]; Fin[Ti]
2283                                        >Sith[Ii1]; Fin[Mi1]
2284                                          >Sith[Ii,y(Ii)], Fin[Mi,y(Mi)]
2285                                               >Sith[Ii,y(Ii),y2(y(Ii))];
2286                                               >Fin[Mi,y(Mi),y2(y(Mi))]
2287                                          >Sith[Ii,y(Ii)], Fin[Mi,y(Mi)] end
2288                                        >Sith[Ii1]; Fin[Mi1] end
2289                                        >Sith[Ii2]; Fin[Mi2]
2290                                          >Sith[Ii,AdBmax(Ii)]; Fin[Mi,AdBmax(Mi)]
2291                                               >Sith[Ii,AdBmax(Ii),y2(AdBmax(Ii))];
2292                                               >Fin[Mi,AdBmax(Mi),y2(AdBmax(Mi))]
2293                                          >Sith[Ii,AdBmax(Ii)]; Fin[Mi,AdBmax(Mi)] end
2294                                          >Sith[C_AdB]
2295                                               >Sith[C_AdB,w_AdB]
2296                                          >Sith[C_AdB] end
2297                                        >Sith[Ii2]; Fin[Mi2] end
2298                              >Sith[Ti]; Fin[Ti] end
2299                         >Sith[T]; Fin[T] end
2300               >structure end
2301               >Konp[Obs5A_AdB_3]
2302                >of Sith[Obs5A_AdB_3], Fin[Obs]
2303                >E[Ц,e[Ц](5A_AdB_3)], E[Ц,e[Ц](R5)]
2304               >Konp[Obs5A_AdB_3]end
2305               >Konr[Descript(T,R5)]
2306                >E[Ч,e[Ч](T)]
2307                >E[Ц,e[Ц](R5)]
2308               >Konr[Descript(T,R5)]end
2309               >Konr[Prior,Sits[R5]]
2310                >Sith[To]
2311                >Sith[Ti]
2312               >Konr[Prior,Sits[R5]]end
2313               >Konr[Prior,Sits[R5]]
2314                >Fin[Mo,x(Mo,p)]
2315                >Fin[Mo,x(Mo,p)+1]
2316               >Konr[Prior,Sits[R5]]end
2317               >Konr[Prior,Sits[R5]]
2318                >Fin[Mi,y(Mi,p)]
2319                >Fin[Mi,y(Mi,p)+1]
2320               >Konr[Prior,Sits[R5]]end
2321             >Sits[5A_AdB_3]end
2322     >Konw[4AdBelt_4]
2323      >of Sits[5A_AdB_3]
2324      >w_AdB1,w_AdB2=groups(m_AdB)
2325      >w_AdB1\=w_AdB2
2326      >Konr[w_AdB1,w_AdB2]
2327        >Sith[C,w_AdB1]
2328        >Sith[C,w_AdB2]
2329      >Konr[w_AdB1,w_AdB2]end
2330      >Sits[G_AdB]=Konf[int](Konr[w_AdB1,w_AdB2])
2331      >E[Ч,e[Ч](Subject_AdB)]=Sith[C,w_AdB1]
2332       >E[Ч,e[Ч](Object_AdB)]=Sith[C,w_AdB2]
2333      >Konw[Causality_AdB]=Class(Konr[w_AdB1,w_AdB2])
2334           >Sits[G_AdB]
2335             >structure
2336                 >Sith[G_AdB]
2337                      >Sith[Subject_AdB]
```

```
2338                                              >Sith[Subject_AdB,d_AdB(Subject_AdB)]
2339                                                  >Sith[Subject_AdB,d_AdB(Subject_AdB),
2340                                                      >c_AdB(d_AdB(Subject_AdB))]
2341                              >Sith[Subject_AdB,d_AdB(Subject_AdB)] end
2342                              >Sith[Subject_AdB] end
2343                              >Sith[Object_AdB]
2344                                      >Sith[Object_AdB,d_AdB(Object_AdB)]
2345                                          >Sith[Object_AdB,d_AdB(Object_AdB),
2346                                              >c_AdB(d_AdB(Object_AdB))]
2347                                      >Sith[Object_AdB,d_AdB(Object_AdB)] end
2348                              >Sith[Object_AdB] end
2349                      >Sith[G_AdB] end
2350              >structure end
2351              >Konw[Causality_AdB]
2352               >of Sith[Subject_AdB], Sith[Subject_AdB,d_AdB(Subject_AdB)],
2353               >Sith[Subject_AdB,d_AdB(Subject_AdB),c_AdB(d_AdB(Subject_AdB))]
2354               >Sith[Object_AdB], Sith[Object_AdB,d_AdB(Object_AdB)],
2355               >Sith[Object_AdB,d_AdB(Object_AdB),c_AdB(d_AdB(Object_AdB))]
2356               >Konw[Causality_AdB]end
2357          >Sits[G_AdB] end
2358      >Sith[C_AdB]=Konw[M](Sits[C_AdB])
2359       >E[Ч,e[Ч](G_AdB)]=E[3,e[3](G_AdB)]
2360      >Sits[5A_AdB_4]
2361  >Konw[4AdBelt_4] end
2362              >Sits[5A_AdB_4]
2363               >structure
2364                      >Sits[R5]
2365                       >E[Ц,e[Ц](R5)]
2366                      >Sith[Obs5A_AdB_4]; Fin[Obs]
2367                      >Sith[T]; Fin[T]
2368                          >Sith[To]; Fin[To]
2369                              >Sith[Io,x(Io)]; Fin[Mo,x(Mo)]
2370                                  >Sith[Io,x(Io),x2(x(Io))];
2371                                  >Fin[Mo,x(Mo),x2(x(Mo))]
2372                              >Sith[Io,x(Io)]; Fin[Mo,x(Mo)] end
2373                          >Sith[To]; Fin[To] end
2374                          >Sith[Ti]; Fin[Ti]
2375                              >Sith[Ii1]; Fin[Mi1]
2376                                  >Sith[Ii,y(Ii)], Fin[Mi,y(Mi)]
2377                                      >Sith[Ii,y(Ii),y2(y(Ii))];
2378                                      >Fin[Mi,y(Mi),y2(y(Mi))]
2379                                  >Sith[Ii,y(Ii)], Fin[Mi,y(Mi)] end
2380                              >Sith[Ii1]; Fin[Mi1] end
2381                              >Sith[Ii2]; Fin[Mi2]
2382                                  >Sith[Ii,AdBmax(Ii)]; Fin[Mi,AdBmax(Mi)]
2383                                      >Sith[Ii,AdBmax(Ii),y2(AdBmax(Ii))];
2384                                      >Fin[Mi,AdBmax(Mi),y2(AdBmax(Mi))]
2385                                  >Sith[Ii,AdBmax(Ii)]; Fin[Mi,AdBmax(Mi)] end
2386                                  >Sith[C_AdB]
2387                                      >Sith[C_AdB,w_AdB]
2388                                  >Sith[C_AdB]end
2389                                  >Sith[G_AdB]
2390                                      >Sith[Subject_AdB]
2391                                          >Sith[Subject_AdB,
2392                                          >d_AdB(Subject_AdB)]
2393                                              >Sith[Subject_AdB,
2394                                              >d_AdB(Subject_AdB),
2395                                              >c_AdB(d_AdB(Subject_AdB))]
2396                                          >Sith[Subject_AdB,
2397                                          >d_AdB(Subject_AdB)] end
```

```
2398                                                      >Sith[Subject_AdB] end
2399                                                      >Sith[Object_AdB]
2400                                                          >Sith[Object_AdB,
2401                                                          >d_AdB(Object_AdB)]
2402                                                              >Sith[Object_AdB,
2403                                                              >d_AdB(Object_AdB),
2404                                                              >c_AdB(d_AdB(Object_AdB))]
2405                                                          >Sith[Object_AdB,
2406                                                          >d_AdB(Object_AdB)] end
2407                                                      >Sith[Object_AdB] end
2408                                              >Sits[G_AdB] end
2409                                      >Sith[Ii2]; Fin[Mi2] end
2410                              >Sith[Ti]; Fin[Ti] end
2411                          >Sith[T]; Fin[T] end
2412              >structure end
2413              >Konw[Conc(C_AdB,G_AdB)]
2414               >E[Ч,e[Ч](C_AdB)]
2415               >E[Ч,e[Ч](G_AdB)]
2416              >Konw[Conc(C_AdB,G_AdB)]end
2417              >Konw[Causality_AdB]
2418               >of Sith[Subject_AdB], Sith[Subject_AdB,d_AdB(Subject_AdB)],
2419               >Sith[Subject_AdB,d_AdB(Subject_AdB),c_AdB(d_AdB(Subject_AdB))]
2420               >Sith[Object_AdB], Sith[Object_AdB,d_AdB(Object_AdB)],
2421               >Sith[Object_AdB,d_AdB(Object_AdB),c_AdB(d_AdB(Object_AdB))]
2422              >Konw[Causality_AdB]end
2423              >Konp[Obs5A_AdB_4]
2424               >of Sith[Obs5A_AdB_4], Fin[Obs]
2425               >E[Ц,e[Ц](5A_AdB_4)], E[Ц,e[Ц](R5)]
2426              >Konp[Obs5A_AdB_4]end
2427              >Konr[Descript((T,G_AdB),R5)]
2428               >E[Ч,e[Ч](T)], E[Ч,e[Ч](G_AdB)]
2429               >E[Ц,e[Ц](R5)]
2430              >Konr[Descript((T,G_AdB),R5)]end
2431              >Konr[Prior,Sits[R5]]
2432               >Sith[To]
2433               >Sith[Ti]
2434              >Konr[Prior,Sits[R5]]end
2435              >Konr[Prior,Sits[R5]]
2436               >Fin[Mo,x(Mo,p)]
2437               >Fin[Mo,x(Mo,p)+1]
2438              >Konr[Prior,Sits[R5]]end
2439              >Konr[Prior,Sits[R5]]
2440               >Fin[Mi,y(Mi,p)]
2441               >Fin[Mi,y(Mi,p)+1]
2442              >Konr[Prior,Sits[R5]]end
2443          >Sits[5A_AdB_4]end
2444      >Konf[Ex_AdB]
2445       >of Sits[5A_AdB_4]
2446      >Konr[Descript(G_AdB,R5)]
2447       >E[Ч,e[Ч](G_AdB)]
2448       >E[Ц,e[Ц](R5)]
2449      >Konr[Descript(G_AdB,R5)]end
2450      >if (Konr[Descript(G_AdB,R5)]=omnivial):
2451       >ExAdBA=1
2452      >else:
2453       >ExAdBA=0
2454      >endif
2455      >i_AdB=1
2456      >ExAdBloop=1
2457      >while (ExAdBloop=1):
```

```
2458        >if (ExAdBA=1):
2459            >Sits[Modeli_AdB]
2460             >structure
2461                 >Sith[chi]; Fin[chi]
2462                     >Sith[Subject_AdB]
2463                         >Sith[Subject_AdB,d_AdB(Subject_AdB)]
2464                             >Sith[Subject_AdB,d_AdB(Subject_AdB),
2465                             >c_AdB(d_AdB(Subject_AdB))]
2466                         >Sith[Subject_AdB,d_AdB(Subject_AdB)] end
2467                     >Sith[Subject_AdB] end
2468                     >Sith[Object_AdB]
2469                         >Sith[Object_AdB,d_AdB(Object_AdB)]
2470                             >Sith[Object_AdB,d_AdB(Object_AdB),
2471                             >c_AdB(d_AdB(Object_AdB))]
2472                         >Sith[Object_AdB,d_AdB(Object_AdB)] end
2473                     >Sith[Object_AdB] end
2474                 >Sith[chi]; Fin[chi] end
2475             >structure end
2476             >Konw[NG_AdB]
2477              >of Sith[Subject_AdB], Sith[Subject_AdB],d_AdB](Subject_AdB)],
2478              >Sith[Subject_AdB],d_AdB](Subject_AdB)),c_AdB](d_AdB](Subject_AdB)))]
2479              >Sith[Object_AdB]], Sith[Object_AdB],d_AdB](Object_AdB)],
2480              >Sith[Object_AdB],d_AdB](Object_AdB)),c_AdB](d_AdB](Object_AdB)))]
2481             >Konw[NG_AdB] end
2482         >Sits[Modeli_AdB]] end
2483         >Sits[h,Ii_AdB]]
2484             >structure
2485                 >Sith[chi]; Fin[chi]
2486                     >Sith[Subject_AdB]
2487                         >Sith[Subject_AdB,d_AdB(Subject_AdB)]
2488                             >Sith[Subject_AdB,d_AdB(Subject_AdB),
2489                             >c_AdB(d_AdB(Subject_AdB))]
2490                         >Sith[Subject_AdB,d_AdB(Subject_AdB)] end
2491                     >Sith[Subject_AdB] end
2492                 >Sith[chi]; Fin[chi] end
2493             >structure end
2494         >Sits[h,Ii_AdB] end
2495     >Class(Sith[chi],Fin[chi](Sits[h,Ii_AdB]))=
2496     >Konf[M](E[Ц,eЦ](Modeli_AdB)],Konw[NG_AdB])
2497     >Konr[Subject_AdB,chi]
2498       >Sith[Subject_AdB]
2499       >Sith[chi]
2500      >Konr[Subject_AdB,chi]end
2501     >Konr[Subject_AdB,chi]=Konf[M](Konw[NG_AdB])
2502     >Class(Konr[Subject_AdB,chi])=Ef
2503     >Sith[Ii_AdB] Sith[Ii_AdB,y_AdB(Ii_AdB)]
2504      >E[Ч^5,e[Ч^5](Ii_AdB)]=Class(Sith[chi],Fin[chi])
2505     >Fin[Mi_AdB], Fin[Mi_AdB,y_AdB(Mi_AdB)]=Konw[M](Sits[Modeli_AdB])
2506      >E[Ч^5,e[Ч^5](Mi_AdB)]=E[3,e[3](Modeli_AdB)]
2507     >q_AdB(Ii_AdB)=abs(Class(Sith[chi],Fin[chi]))
2508     >q_AdB(Mi_AdB)=abs(ALL(E[3,e[3](Model_AdB)]))
2509     >Sits[5A_AdB_5]
2510     >elif (ExAdBA=0):
2511     >s_AdB[1:i_AdB]
2512     >w_AdB1(i_AdB)=w_AdB1
2513     >w_AdB2(i_AdB)=w_AdB2
2514     >Sith[i_AdB,Subject_AdB(i_AdB),d(Subject_AdB(i_AdB)),
2515     >c_AdB(d_AdB(Subject_AdB(i_AdB)))]=
2516     >Sith[Subject_AdB,d_AdB(Subject_AdB),
2517     >c_AdB(d_AdB(Subject_AdB))]
```

42

```
2518        >Sith[i_AdB,Subject_AdB,d_AdB(Subject(i_AdB))]=
2519        >Sith[Subject_AdB,d_AdB(Subject_AdB)]
2520        >Sith[i_AdB,Subject_AdB(i_AdB)]=Sith[Subject_AdB]
2521        >Sith[i_AdB,Object_AdB(i_AdB),d(Object_AdB(i_AdB)),
2522        >c_AdB(d_AdB(Object_AdB(i_AdB)))]=
2523        >Sith[Object_AdB,d_AdB(Object_AdB),c_AdB(d_AdB(Object_AdB))]
2524        >Sith[i_AdB,Object_AdB,d_AdB(Object(i_AdB))]=Sith[Object_AdB,d_AdB(Object_AdB)]
2525        >Sith[i_AdB,Object_AdB(i_AdB)]=Sith[Object_AdB]
2526            >Sits[F(i_AdB)]
2527             >structure
2528                    >Sith[Guess,s_AdB]
2529                        >Sith[s_AdB,Subject(s_AdB)]
2530                            >Sith[s_AdB,Subject_AdB(s_AdB),d_AdB(Subject(s_AdB))]
2531                                >Sith[s_AdB,Subject_AdB(s_AdB),
2532                                >d_AdB(Subject_AdB(s_AdB)),c_AdB(d_AdB(Subject_Ad
2533                                B(s_AdB)))]
2534                            >Sith[s_AdB,Subject_AdB(s_AdB),
2535                            >d_AdB(Subject_AdB(s_AdB))] end
2536                        >Sith[s_AdB,Subject(s_AdB)] end
2537                        >Sith[s_AdB,Object(s_AdB)]
2538                            >Sith[s_AdB,Object_AdB(s_AdB),d_AdB(Object_AdB(s_AdB))]
2539                                    >Sith[s_AdB,Object_AdB(s_AdB),
2540                                    >d_AdB(Object_AdB(s_AdB)),
2541                                    >c_AdB(d_AdB(Object_AdB(s_AdB)))]
2542                            >Sith[s_AdB,Object_AdB(s_AdB),
2543                            >d_AdB(Object_AdB(s_AdB))] end
2544                        >Sith[s_AdB,Object_AdB(s_AdB)] end
2545                    >Sith[Guess,s_AdB] end
2546             >structure end
2547             >Konw[Causality_AdB]
2548              >of Sith[s_AdB,Subject_AdB(s_AdB)], Sith[s_AdB,Subject_AdB(s_AdB),
2549              >d_AdB(Subject_AdB(s_AdB))],
2550              >Sith[Subject_AdB(s_AdB),d_AdB(Subject_AdB(s_AdB)),
2551              >c_AdB(d_AdB(Subject_AdB(s_AdB)))]
2552              >Sith[s_AdB,Object_AdB(s_AdB)], Sith[s_AdB,Object_AdB(s_AdB),
2553              >d_AdB(Object_AdB(s_AdB))],
2554              >Sith[s_AdB,Object_AdB(s_AdB),
2555              >d_AdB(Object_AdB(s_AdB)),c_AdB(d_AdB(Object_AdB(s_AdB)))]
2556             >Konw[Causality_AdB]end
2557            >Sits[F(i_AdB)]end
2558        >w_AdB1,w_AdB2=groups(m_AdB)
2559         >w_AdB1\=w_AdB1(s_AdB)
2560         >w_AdB2\=w_AdB2(s_AdB)
2561        >Konr[w_AdB1,w_AdB2]
2562          >Sith[C,w_AdB1]
2563          >Sith[C,w_AdB2]
2564         >Konr[w_AdB1,w_AdB2]end
2565         >Sits[G]=Konf[int](Konr[w_AdB1,w_AdB2])
2566         >E[Ч,e[Ч](Subject_AdB)]=Sith[C,w_AdB1]
2567          >E[Ч,e[Ч](Object_AdB)]=Sith[C,w_AdB2]
2568        >Konw[Causality_AdB]=Class(Konr[w_AdB1,w_AdB2])
2569        >Sits[G_AdB]
2570         >E[Ц,e[Ц](G_AdB)]\=E[Ц,e[Ц](F(i_AdB))]
2571        >if (abs(Sits[G_AdB])\=0):
2572         >ExAdBloop=1
2573        >elif (abs(Sits[G_AdB])=0):
2574         >ExAdBloop=0
2575         >proceed
2576        >endif
2577            >Sits[G_AdB]
```

```
2578              >structure
2579                  >Sith[Subject_AdB]
2580                      >Sith[Subject_AdB,d_AdB(Subject_AdB)]
2581                          >Sith[Subject_AdB,d_AdB(Subject_AdB),
2582                          >c_AdB(d_AdB(Subject_AdB))]
2583                      >Sith[Subject_AdB,d_AdB(Subject_AdB)] end
2584                  >Sith[Subject_AdB] end
2585                  >Sith[Object_AdB]
2586                      >Sith[Object_AdB,d_AdB(Object_AdB)]
2587                          >Sith[Object_AdB,d_AdB(Object_AdB),
2588                          >c_AdB(d_AdB(Object_AdB))]
2589                      >Sith[Object_AdB,d_AdB(Object_AdB)] end
2590                  >Sith[Object_AdB] end
2591              >structure end
2592              >Konw[Causality_AdB]
2593               >of Sith[Subject_AdB], Sith[Subject_AdB,d_AdB(Subject_AdB)],
2594               >Sith[Subject_AdB,d_AdB(Subject_AdB),c_AdB(d_AdB(Subject_AdB))]
2595               >Sith[Object_AdB], Sith[Object_AdB,d_AdB(Object_AdB)],
2596               >Sith[Object_AdB,d_AdB(Object_AdB),c_AdB(d_AdB(Object_AdB))]
2597              >Konw[Causality_AdB]end
2598          >Sits[G_AdB] end
2599      >Sith[G_AdB]=Konw[M](Sits[G_AdB])
2600       >E[Ч,e[Ч](G_AdB)]=E[3,e[3](G_AdB)]
2601      >Kont[AdB(i_AdB)]
2602      >i_AdB=i_AdB+1
2603    >endif
2604   >else
2605    >Sits[5A_AdB_1]
2606    >Sits[5A_AdB]=Sits[5A_AdB_1]
2607     >E[Ц,e[Ц](5A_AdB)]=E[Ц,e[Ц](5A_AdB_1)]
2608    >endloop
2609  >Konf[Ex_AdB]end
2610          >Sits[5A_AdB_5]
2611           >structure
2612              >Sits[R5]
2613               >E[Ц,e[Ц](R5)]
2614              >Sith[Obs5A_AdB_5]; Fin[Obs]
2615              >Sith[T]; Fin[T]
2616                  >Sith[To]; Fin[To]
2617                      >Sith[Io,x(Io)]; Fin[Mo,x(Mo)]
2618                          >Sith[Io,x(Io),x2(x(Io))];
2619                          >Fin[Mo,x(Mo),x2(x(Mo))]
2620                      >Sith[Io,x(Io)]; Fin[Mo,x(Mo)] end
2621                  >Sith[To]; Fin[To] end
2622                  >Sith[Ti]; Fin[Ti]
2623                      >Sith[Ii1]; Fin[Mi1]
2624                          >Sith[Ii,y(Ii)], Fin[Mi,y(Mi)]
2625                              >Sith[Ii,y(Ii),y2(y(Ii))];
2626                              >Fin[Mi,y(Mi),y2(y(Mi))]
2627                          >Sith[Ii,y(Ii)], Fin[Mi,y(Mi)] end
2628                      >Sith[Ii2]
2629                          >Sith[Ii,AdBmax(Ii)]; Fin[Mi,AdBmax(Mi)]
2630                              >Sith[Ii,AdBmax(Ii),
2631                              >y2(AdBmax(Ii))];
2632                              >Fin[Mi,AdBmax(Mi),y2(AdBmax(Mi))]
2633                          >Sith[Ii,AdBmax(Ii)];
2634                          >Fin[Mi,AdBmax(Mi)] end
2635                      >Sith[Ii_AdB]; Fin[Mi_AdB]
2636                          >Sith[Ii_AdB,y_AdB(Mi_AdB)];
2637                          >Fin[Mi_AdB,y_AdB(Mi_AdB)]
```

```
2638                                                    >Sith[Ii_AdB]; Fin[Mi_AdB] end
2639                                                 >Sith[Ii2] end
2640                                          >Sith[Ii1]; Fin[Mi1] end
2641                                    >Sith[Ti]; Fin[Ti] end
2642                              >Sith[T]; Fin[T] end
2643                  >structure end
2644                  >Konp[Obs5A_AdB_5]
2645                   >of Sith[Obs5A_AdB_5], Fin[Obs]
2646                   >E[Ц,е[Ц](5A_AdB_5)], E[Ц,е[Ц](R5)]
2647                  >Konp[Obs5A_AdB_5]end
2648                  >Konr[Descript(T,R5)]
2649                   >E[Ч,е[Ч](T)]
2650                   >E[Ц,е[Ц](R5)]
2651                  >Konr[Descript(T,R5)]end
2652                  >Konr[Prior,Sits[R5]]
2653                   >Sith[To]
2654                   >Sith[Ti]
2655                  >Konr[Prior,Sits[R5]]end
2656                  >Konr[Prior,Sits[R5]]
2657                   >Fin[Mo,x(Mo,p)]
2658                   >Fin[Mo,x(Mo,p)+1]
2659                  >Konr[Prior,Sits[R5]]end
2660                  >Konr[Prior,Sits[R5]]
2661                   >Fin[Mi,y(Mi,p)]
2662                   >Fin[Mi,y(Mi,p)+1]
2663                  >Konr[Prior,Sits[R5]]end
2664                 >Sits[5A_AdB_5]end
2665          >Konw[5A_AdB_6]
2666           >of Sits[5A_AdB_5]
2667                  >Sits[Ti2]
2668                   >structure
2669                      >Sits[Ii2,new]
2670                           >SithSith[Ii,AdBmax(Ii),y2(AdBmax(Ii))];
2671                           >Fin[Mi,AdBmax(Mi),y2(AdBmax(Mi))]
2672                           >Sith[Ii_AdB,y_AdB(Mi_AdB)]; Fin[Mi_AdB,y_AdB(Mi_AdB)]
2673                      >Sits[Ii2,new] end
2674                   >structure end
2675                  >Sits[Ti2] end
2676       >Sith[Mi2new,y(Mi2new)]=Fin[Mi,AdBmax(Mi),y2(AdBmax(Mi))]+Fin[Mi_AdB,y_AdB(Mi_AdB)]
2677       >\Fin[Mi2]
2678       >Fin[Mi2]=Fin[Mi2new]
2679        >E[Ч,е[Ч](Mi2)]=E[Ч,е[Ч](Mi2new)]
2680       >Sits[5A_AdB_6]
2681      >Konw[5A_AdB_6] end
2682              >Sits[5A_AdB_6]
2683               >structure
2684                   >Sits[R5]
2685                    >E[Ц,е[Ц](R5)]
2686                   >Sith[Obs5A_AdB_6]; Fin[Obs]
2687                   >Sith[T]; Fin[T]
2688                        >Sith[To]; Fin[To]
2689                            >Sith[Io,x(Io)]; Fin[Mo,x(Mo)]
2690                                >Sith[Io,x(Io),x2(x(Io))];
2691                                >Fin[Mo,x(Mo),x2(x(Mo))]
2692                            >Sith[Io,x(Io)]; Fin[Mo,x(Mo)] end
2693                        >Sith[To]; Fin[To] end
2694                        >Sith[Ti]; Fin[Ti]
2695                            >Sith[Ii1]; Fin[Mi1]
2696                                >Sith[Ii,y(Ii)], Fin[Mi,y(Mi)]
```

```
2697                                               >Sith[Ii,y(Ii),y2(y(Ii))];
2698                                               >Fin[Mi,y(Mi),y2(y(Mi))]
2699                                        >Sith[Ii,y(Ii)], Fin[Mi,y(Mi)] end
2700                                        >Sith[Ii2]; Fin[Mi2]
2701                                               >Sith[Ii2,y(Ii2)]; Fin[Mi2,y(Mi2)]
2702                                        >Sith[Ii2]; Fin[Mi2] end
2703                                 >Sith[Ii1]; Fin[Mi1] end
2704                          >Sith[Ti]; Fin[Ti] end
2705                     >Sith[T]; Fin[T] end
2706              >structure end
2707              >Konp[Obs5A_AdB_6]
2708               >of Sith[Obs5A_AdB_6], Fin[Obs]
2709               >E[Ц,e[Ц](5A_AdB_6)], E[Ц,e[Ц](R5)]
2710              >Konp[Obs5A_AdB_6]end
2711              >Konr[Descript(T,R5)]
2712               >E['I,e['I](T)]
2713               >E[Ц,e[Ц](R5)]
2714              >Konr[Descript(T,R5)]end
2715              >Konr[Prior,Sits[R5]]
2716               >Sith[To]
2717               >Sith[Ti]
2718              >Konr[Prior,Sits[R5]]end
2719              >Konr[Prior,Sits[R5]]
2720               >Fin[Mo,x(Mo,p)]
2721               >Fin[Mo,x(Mo,p)+1]
2722              >Konr[Prior,Sits[R5]]end
2723              >Konr[Prior,Sits[R5]]
2724               >Fin[Mi,y(Mi,p)]
2725               >Fin[Mi,y(Mi,p)+1]
2726              >Konr[Prior,Sits[R5]]end
2727              >Sits[5A_AdB_6]end
2728      >Konw[5A_AdB_7]
2729       >of Sits[5A_AdB_6]
2730      >Fin[Mi,y(Mi),y2(y(Mi))]=Fin[Mi,y(Mi),y2(y(Mi))]+Fin[Mi2,y(Mi2)]
2731      >Fin[Mi,y(Mi)]=Fin[Mi,y(Mi)]+Fin[Mi2]
2732      >q(Mi)=q(Mi)-1
2733      >y(Mi)[1:q(Mi)]
2734      >y(Mi,p)[1:q(Mi)-1]
2735      >Sits[5A_AdB_7]
2736      >Konw[5A_AdB_7] end
2737              >Sits[5A_AdB_7]
2738               >structure
2739                  >Sits[R5]
2740                   >E[Ц,e[Ц](R5)]
2741                  >Sith[Obs5A_AdB_7]; Fin[Obs]
2742                  >Sith[T]; Fin[T]
2743                     >Sith[To]; Fin[To]
2744                          >Sith[Io,x(Io)]; Fin[Mo,x(Mo)]
2745                               >Sith[Io,x(Io),x2(x(Io))];
2746                               >Fin[Mo,x(Mo),x2(x(Mo))]
2747                          >Sith[Io,x(Io)]; Fin[Mo,x(Mo)] end
2748                     >Sith[To]; Fin[To] end
2749                     >Sith[Ti]; Fin[Ti]
2750                          >Sith[Ii,y(Ii)], Fin[Mi,y(Mi)]
2751                               >Sith[Ii,y(Ii),y2(y(Ii))];
2752                               >Fin[Mi,y(Mi),y2(y(Mi))]
2753                          >Sith[Ii,y(Ii)], Fin[Mi,y(Mi)] end
2754                     >Sith[Ti]; Fin[Ti] end
2755                  >Sith[T]; Fin[T] end
2756               >structure end
```

46

```
2757              >Konp[Obs5A_AdB_7]
2758               >of Sith[Obs5A_AdB_7], Fin[Obs]
2759               >E[Ц,e[Ц](5A_AdB_7)], E[Ц,e[Ц](R5)]
2760              >Konp[Obs5A_AdB_6]end
2761              >Konr[Descript(T,R5)]
2762               >E[Ч,e[Ч](T)]
2763               >E[Ц,e[Ц](R5)]
2764              >Konr[Descript(T,R5)]end
2765              >Konr[Prior,Sits[R5]]
2766               >Sith[To]
2767               >Sith[Ti]
2768              >Konr[Prior,Sits[R5]]end
2769              >Konr[Prior,Sits[R5]]
2770               >Fin[Mo,x(Mo,p)]
2771               >Fin[Mo,x(Mo,p)+1]
2772              >Konr[Prior,Sits[R5]]end
2773              >Konr[Prior,Sits[R5]]
2774               >Fin[Mi,y(Mi,p)]
2775               >Fin[Mi,y(Mi,p)+1]
2776              >Konr[Prior,Sits[R5]]end
2777             >Sits[5A_AdB_7]end
2778        >Sits[5A_AdB]=Sits[5A_AdB_7]
2779          >E[Ц,e[Ц](5A_AdB)]=E[Ц,e[Ц](5A_AdB_7)]
2780       >Kont[AdBelt]end
2781       >Kont[AdB(i_AdB)]
2782             >Sits[5A_AdB_4(i_AdB)]
2783              >structure
2784                   >Sits[R5]
2785                    >E[Ц,e[Ц](R5)]
2786                   >Sith[Obs5A_AdB-4(i_AdB)]; Fin[Obs]
2787                   >Sith[T]; Fin[T]
2788                       >Sith[To]; Fin[To]
2789                           >Sith[Io,x(Io)]; Fin[Mo,x(Mo)]
2790                               >Sith[Io,x(Io),x2(x(Io))];
2791                               >Fin[Mo,x(Mo),x2(x(Mo))]
2792                           >Sith[Io,x(Io)]; Fin[Mo,x(Mo)] end
2793                       >Sith[To]; Fin[To] end
2794                       >Sith[Ti]; Fin[Ti]
2795                           >Sith[Ii1]; Fin[Mi1]
2796                               >Sith[Ii,y(Ii)], Fin[Mi,y(Mi)]
2797                                   >Sith[Ii,y(Ii),y2(y(Ii))];
2798                                   >Fin[Mi,y(Mi),y2(y(Mi))]
2799                               >Sith[Ii,y(Ii)], Fin[Mi,y(Mi)] end
2800                           >Sith[Ii1]; Fin[Mi1] end
2801                           >Sith[Ii2]; Fin[Mi2]
2802                               >Sith[Ii,AdBmax(Ii)]; Fin[Mi,AdBmax(Mi)]
2803                                   >Sith[Ii,AdBmax(Ii),y2(AdBmax(Ii))];
2804                                   >Fin[Mi,AdBmax(Mi),y2(AdBmax(Mi))]
2805                               >Sith[Ii,AdBmax(Ii)]; Fin[Mi,AdBmax(Mi)] end
2806                           >Sith[C_AdB]
2807                                >Sith[C_AdB,w_AdB]
2808                           >Sith[C_AdB]end
2809                           >Sith[G_AdB]
2810                               >Sith[Subject_AdB]
2811                                   >Sith[Subject_AdB,
2812                                   >d_AdB(Subject_AdB)]
2813                                       >Sith[Subject_AdB,
2814                                       >d_AdB(Subject_AdB),
2815                                       >c_AdB(d_AdB(Subject_AdB))]
```

```
2816                                                      >Sith[Subject_AdB,
2817                                                       >d_AdB(Subject_AdB)] end
2818                                              >Sith[Subject_AdB] end
2819                                             >Sith[Object_AdB]
2820                                                  >Sith[Object_AdB,
2821                                                   >d_AdB(Object_AdB)]
2822                                                      >Sith[Object_AdB,
2823                                                       >d_AdB(Object_AdB),
2824                                                        >c_AdB(d_AdB(Object_AdB))]
2825                                                  >Sith[Object_AdB,
2826                                                   >d_AdB(Object_AdB)] end
2827                                             >Sith[Object_AdB] end
2828                                        >Sith[G_AdB] end
2829                                  >Sith[Ii2]; Fin[Mi2] end
2830                           >Sith[Ti]; Fin[Ti] end
2831                     >Sith[T]; Fin[T] end
2832               >structure end
2833               >Konw[Conc(C_AdB,G_AdB)]
2834                >E[Ч,e[Ч](C_AdB)]
2835                >E[Ч,e[Ч](G_AdB)]
2836               >Konw[Conc(C_AdB,G_AdB)]end
2837               >Konw[Causality_AdB]
2838                >of Sith[Subject_AdB], Sith[Subject_AdB,d_AdB(Subject_AdB)],
2839                >Sith[Subject_AdB,d_AdB(Subject_AdB),c_AdB(d_AdB(Subject_AdB))]
2840                >Sith[Object_AdB], Sith[Object_AdB,d_AdB(Object_AdB)],
2841                >Sith[Object_AdB,d_AdB(Object_AdB),c_AdB(d_AdB(Object_AdB))]
2842               >Konw[Causality_AdB]end
2843               >Konp[Obs5A_AdB_4]
2844                >of Sith[Obs5A_AdB_4(i_AdB)], Fin[Obs]
2845                >E[Ц,e[Ц](5A_AdB_4)(i_AdB)], E[Ц,e[Ц](R5)]
2846               >Konp[Obs5A_AdB_4(i_AdB)]end
2847               >Konr[Descript((T,G_AdB),R5)]
2848                >E[Ч,e[Ч](T)], E[Ч,e[Ч](G_AdB)]
2849                >E[Ц,e[Ц](R5)]
2850               >Konr[Descript((T,G_AdB),R5)]end
2851               >Konr[Prior,Sits[R5]]
2852                >Sith[To]
2853                >Sith[Ti]
2854               >Konr[Prior,Sits[R5]]end
2855               >Konr[Prior,Sits[R5]]
2856                >Fin[Mo,x(Mo,p)]
2857                >Fin[Mo,x(Mo,p)+1]
2858               >Konr[Prior,Sits[R5]]end
2859               >Konr[Prior,Sits[R5]]
2860                >Fin[Mi,y(Mi,p)]
2861                >Fin[Mi,y(Mi,p)+1]
2862               >Konr[Prior,Sits[R5]]end
2863             >Sits[5A_AdB_4]end
2864     >Kont[AdB(i_AdB)]end
2865     >Kont[ReBelt]
2866      >Sits[5A_AdB_1]=Sits[5A(l[0])]
2867       >E[Ц,e[Ц](5A_AdB_1)]=E[Ц,e[Ц](5A(l[0]))]
2868            >Sits[5A_ReB_1]
2869             >structure
2870                 >Sits[R5]
2871                  >E[Ц,e[Ц](R5)]
2872                 >Sith[Obs5A_ReB_1]; Fin[Obs]
2873                 >Sith[T]; Fin[T]
2874                     >Sith[To]; Fin[To]
2875                         >Sith[Io,x(Io)]; Fin[Mo,x(Mo)]
```

```
2876                                              >Sith[Io,x(Io),x2(x(Io))];
2877                                              >Fin[Mo,x(Mo),x2(x(Mo))]
2878                                    >Sith[Io,x(Io)]; Fin[Mo,x(Mo)] end
2879                          >Sith[To]; Fin[To] end
2880                          >Sith[Ti]; Fin[Ti]
2881                                    >Sith[Ii,y(Ii)], Fin[Mi,y(Mi)]
2882                                              >Sith[Ii,y(Ii),y2(y(Ii))];
2883                                              >Fin[Mi,y(Mi),y2(y(Mi))]
2884                                    >Sith[Ii,y(Ii)], Fin[Mi,y(Mi)] end
2885                          >Sith[Ti]; Fin[Ti]
2886                >Sith[T]; Fin[T] end
2887          >structure end
2888          >Konp[Obs5A_ReB_1]
2889           >of Sith[Obs5A_ReB_1], Fin[Obs]
2890           >Е[Ц,e[Ц](5A_ReB_1)], E[Ц,e[Ц](R5)]
2891          >Konp[Obs5A_ReB_1]end
2892          >Konr[Descript(T,R5)]
2893           >E[Ч,e[Ч](T)]
2894           >E[Ц,e[Ц](R5)]
2895          >Konr[Descript(T,R5)]end
2896          >Konr[Prior,Sits[R5]]
2897           >Sith[To]
2898           >Sith[Ti]
2899          >Konr[Prior,Sits[R5]]end
2900          >Konr[Prior,Sits[R5]]
2901           >Fin[Mo,x(Mo,p)]
2902           >Fin[Mo,x(Mo,p)+1]
2903          >Konr[Prior,Sits[R5]]end
2904          >Konr[Prior,Sits[R5]]
2905           >Fin[Mi,y(Mi,p)]
2906           >Fin[Mi,y(Mi,p)+1]
2907          >Konr[Prior,Sits[R5]]end
2908         >Sits[5A_ReB_1]end
2909   >Konw[4ReBelt_2]
2910    >of Sits[5A_NiB_1]
2911    >Sits[5A_NiB_2]
2912    >y(Mi)[1:q(Mi)-1]
2913    >y(Mi,p)[1:q(Mi)-2]
2914    >Fin[Mi,y(Mi)]
2915     >\Fin[Mi,y(Mi,p)max+1]
2916    >q(Mi)=q(Mi)-1
2917    >y(Mi)[1:q(Mi)]
2918    >y(Mi,p)[1:q(Mi)-1]
2919   >Konw[4ReBelt_2]end
2920          >Sits[5A_ReB_2]
2921           >structure
2922              >Sits[R5]
2923               >E[Ц,e[Ц](R5)]
2924              >Sith[Obs5A_ReB_2]; Fin[Obs]
2925              >Sith[T]; Fin[T]
2926                    >Sith[To]; Fin[To]
2927                          >Sith[Io,x(Io)]; Fin[Mo,x(Mo)]
2928                                    >Sith[Io,x(Io),x2(x(Io))];
2929                                    >Fin[Mo,x(Mo),x2(x(Mo))]
2930                          >Sith[Io,x(Io)]; Fin[Mo,x(Mo)] end
2931                    >Sith[To]; Fin[To] end
2932                    >Sith[Ti]; Fin[Ti]
2933                          >Sith[Ii,y(Ii)], Fin[Mi,y(Mi)]
2934                                    >Sith[Ii,y(Ii),y2(y(Ii))];
2935                                    >Fin[Mi,y(Mi),y2(y(Mi))]
```

```
2936                                              >Sith[Ii,y(Ii)], Fin[Mi,y(Mi)] end
2937                                      >Sith[Ti]; Fin[Ti]
2938                              >Sith[T]; Fin[T] end
2939                      >structure end
2940                      >Konp[Obs5A_Re_2]
2941                       >of Sith[Obs5A_ReB_2], Fin[Obs]
2942                       >E[Ц,e[Ц](5A_ReB_2)], E[Ц,e[Ц](R5)]
2943                      >Konp[Obs5A_ReB_2]end
2944                      >Konr[Descript(T,R5)]
2945                       >E[Ч,e[Ч](T)]
2946                       >E[Ц,e[Ц](R5)]
2947                      >Konr[Descript(T,R5)]end
2948                      >Konr[Prior,Sits[R5]]
2949                       >Sith[To]
2950                       >Sith[Ti]
2951                      >Konr[Prior,Sits[R5]]end
2952                      >Konr[Prior,Sits[R5]]
2953                       >Fin[Mo,x(Mo,p)]
2954                       >Fin[Mo,x(Mo,p)+1]
2955                      >Konr[Prior,Sits[R5]]end
2956                      >Konr[Prior,Sits[R5]]
2957                       >Fin[Mi,y(Mi,p)]
2958                       >Fin[Mi,y(Mi,p)+1]
2959                      >Konr[Prior,Sits[R5]]end
2960                  >Sits[5A_ReB]end
2961          >Konw[4ReBelt_3]
2962           >of Sits[5A_ReB_2]
2963           >Sith[C1_ReB,c_ReB]=mutual(Sith[R5,a_ReB,b_ReB(a_ReB)])
2964           >while (c_ReB=0):
2965            >\Sits[R5]
2966           >Sits[R5new]\=Sits[R5]
2967                  >Sits[R5new]
2968                   >structure
2969                      >Sith[R5new]; Fin[R5new]
2970                          >Sith[R5new,a_ReBnew]; Fin[R5new,a_ReBnew]
2971                              >Sith[R5new,a_ReBnew,b_ReBnew(a_ReBnew)]
2972                          >Sith[R5new,a_ReBnew]; Fin[R5new,a_ReBnew] end
2973                      >Sith[R5new]; Fin[R5new] end
2974                   >strucutre end
2975                  >Sit[R5new] end
2976            >Sits[R5]=Sits[R5new]
2977             >E[Ц,e[Ц](R5)=E[Ц,e[Ц](R5new)
2978           >else:
2979                  >Sits[C1_ReB]
2980                   >structure
2981                      >Sith[C1_ReB]
2982                          >Sith[C1_ReB,c_ReB]
2983                      >Sith[C1_ReB] end
2984                   >structure end
2985                  >Sits[C1_ReB] end
2986            >Sith[d_ReB]=Classes(Sith[C1_ReB])
2987                  >Sits[C2_ReB]
2988                   >structure
2989                      >Sith[d_ReB]
2990                              >Sith[d_ReB,c_ReB(d_ReB)]
2991                      >Sith[d_ReB] end
2992                   >structure end
2993                  >Sits[C2_ReB] end
2994           >Sith[C_ReB], Fin[C_ReB]=Konw[M](Sits[C2_ReB])
2995            >E[Ч,e[Ч](C_ReB)]=E[3,e[3](C2_ReB)]
```

```
2996          >Sits[5A_ReB_3]
2997         >endloop
2998       >Konw[4ReBelt_3]end
2999            >Sits[5A_ReB_3]
3000             >structure
3001                   >Sits[R5]
3002                    >E[Ц,e[Ц](R5)]
3003                   >Sith[Obs5A_ReB_3]; Fin[Obs]
3004                   >Sith[T]; Fin[T]
3005                      >Sith[To]; Fin[To]
3006                         >Sith[Io,x(Io)]; Fin[Mo,x(Mo)]
3007                            >Sith[Io,x(Io),x2(x(Io))];
3008                            >Fin[Mo,x(Mo),x2(x(Mo))]
3009                         >Sith[Io,x(Io)]; Fin[Mo,x(Mo)] end
3010                      >Sith[To]; Fin[To] end
3011                      >Sith[Ti]; Fin[Ti]
3012                         >Sith[Ii1]; Fin[Mi1]
3013                            >Sith[Ii,y(Ii)]; Fin[Mi,y(Mi)]
3014                               >Sith[Ii,y(Ii),y2(y(Ii))];
3015                               >Fin[Mi,y(Mi),y2(y(Mi))]
3016                            >Sith[Ii,y(Ii)]; Fin[Mi,y(Mi)] end
3017                         >Sith[Ii1]; Fin[Mi1] end
3018                         >Sith[Ii2]; Fin[Mi2]
3019                            >Sith[C_ReB]
3020                               >Sith[C_ReB,w_ReB]
3021                            >Sith[C_ReB] end
3022                         >Sith[Ii2]; Fin[Mi2] end
3023                      >Sith[Ti]; Fin[Ti] end
3024                   >Sith[T]; Fin[T] end
3025             >structure end
3026             >Konp[Obs5A_ReB_3]
3027              >of Sith[Obs5A_ReB_3], Fin[Obs]
3028              >E[Ц,e[Ц](5A_ReB_3)], E[Ц,e[Ц](R5)]
3029             >Konp[Obs5A_Re_3]end
3030             >Konr[Descript(T,R5)]
3031              >E[Ч,e[Ч](T)]
3032              >E[Ц,e[Ц](R5)]
3033             >Konr[Descript(T,R5)]end
3034             >Konr[Prior,Sits[R5]]
3035              >Sith[To]
3036              >Sith[Ti]
3037             >Konr[Prior,Sits[R5]]end
3038             >Konr[Prior,Sits[R5]]
3039              >Fin[Mo,x(Mo,p)]
3040              >Fin[Mo,x(Mo,p)+1]
3041             >Konr[Prior,Sits[R5]]end
3042             >Konr[Prior,Sits[R5]]
3043              >Fin[Mi,y(Mi,p)]
3044              >Fin[Mi,y(Mi,p)+1]
3045             >Konr[Prior,Sits[R5]]end
3046           >Sits[5A_AdB_3]end
3047       >Konw[4ReBelt_4]
3048        >of Sits[5A_ReB_3]
3049         >w_ReB1,w_ReB2=groups(w_ReB)
3050        >w_ReB1\=w_ReB2
3051        >Konr[w_ReB1,w_ReB2]
3052          >Sith[C,w_ReB1]
3053          >Sith[C,w_ReB2]
3054         >Konr[w_ReB1,w_ReB2]end
3055         >Sits[G_ReB]=Konf[int](Konr[w_ReB1,w_ReB2])
```

```
3056        >E[Ч,e[Ч](Subject_ReB)]=Sith[C,w_ReB1]
3057         >E[Ч,e[Ч](Object_ReB)]=Sith[C,w_ReB2]
3058        >Konw[Causality_ReB]=Class(Konr[w_ReB1,w_ReB2])
3059            >Sits[G_ReB]
3060             >structure
3061                >Sith[G_ReB]
3062                    >Sith[Subject_ReB]
3063                        >Sith[Subject_ReB,d_ReB(Subject_ReB)]
3064                            >Sith[Subject_ReB,d_ReB(Subject_ReB),
3065                            >c_ReB(d_ReB(Subject_ReB))]
3066                        >Sith[Subject_ReB,d_ReB(Subject_ReB)] end
3067                    >Sith[Subject_ReB] end
3068                    >Sith[Object_ReB]
3069                        >Sith[Object_ReB,d_ReB(Object_ReB)]
3070                            >Sith[Object_ReB,d_ReB(Object_ReB),
3071                            >c_ReB(d_ReB(Object_ReB))]
3072                        >Sith[Object_ReB,d_ReB(Object_ReB)] end
3073                    >Sith[Object_ReB] end
3074                >Sith[G_ReB] end
3075             >structure end
3076             >Konw[Causality_ReB]
3077              >of Sith[Subject_ReB], Sith[Subject_ReB,d_ReB(Subject_ReB)],
3078              >Sith[Subject_ReB,d_ReB(Subject_ReB),c_ReB(d_ReB(Subject_ReB))]
3079              >Sith[Object_ReB], Sith[Object_ReB,d_ReB(Object_ReB)],
3080              >Sith[Object_ReB,d_ReB(Object_ReB),c_ReB(d_ReB(Object_ReB))]
3081             >Konw[Causality_ReB]end
3082            >Sits[G_ReB] end
3083     >Sith[G_ReB]=Konw[M](Sits[G_ReB])
3084      >E[Ч,e[Ч](G_ReB)]=E[3,e[3](G_ReB)]
3085     >Sits[5A_ReB_4]
3086    >Konw[4ReBelt_4] end
3087            >Sits[5A_ReB_4]
3088             >structure
3089                >Sits[R5]
3090                 >E[Ц,e[Ц](R5)]
3091                >Sith[Obs5A_ReB_4]; Fin[Obs]
3092                >Sith[T]; Fin[T]
3093                    >Sith[To]; Fin[To]
3094                        >Sith[Io,x(Io)]; Fin[Mo,x(Mo)]
3095                            >Sith[Io,x(Io),x2(x(Io))];
3096                            >Fin[Mo,x(Mo),x2(x(Mo))]
3097                        >Sith[Io,x(Io)]; Fin[Mo,x(Mo)] end
3098                    >Sith[To]; Fin[To] end
3099                    >Sith[Ti]; Fin[Ti]
3100                        >Sith[Ii1]; Fin[Mi1]
3101                            >Sith[Ii,y(Ii)]; Fin[Mi,y(Mi)]
3102                                >Sith[Ii,y(Ii),y2(y(Ii))];
3103                                >Fin[Mi,y(Mi),y2(y(Mi))]
3104                            >Sith[Ii,y(Ii)]; Fin[Mi,y(Mi)] end
3105                        >Sith[Ii1]; Fin[Mi1] end
3106                        >Sith[Ii2]; Fin[Mi2]
3107                            >Sith[C_ReB]
3108                                >Sith[C_ReB,w_ReB]
3109                            >Sith[C_ReB] end
3110                            >Sith[G_ReB]
3111                                >Sith[Subject_ReB]
3112                                    >Sith[Subject_ReB,
3113                                    >d_ReB(Subject_ReB)]
```

```
3114                                                           >Sith[Subject_ReB,
3115                                                            >d_ReB(Subject_ReB),
3116                                                             >c_ReB(d_ReB(Subject_ReB))]
3117                                                       >Sith[Subject_ReB,
3118                                                        >d_ReB(Subject_ReB)] end
3119                                                   >Sith[Subject_ReB] end
3120                                                   >Sith[Object_ReB]
3121                                                       >Sith[Object_ReB,
3122                                                        >d_ReB(Object_ReB)]
3123                                                           >Sith[Object_ReB,
3124                                                            >d_ReB(Object_ReB),
3125                                                             >c_ReB(d_ReB(Object_ReB))]
3126                                                       >Sith[Object_ReB,
3127                                                        >d_ReB(Object_ReB)] end
3128                                                   >Sith[Object_ReB] end
3129                                               >Sits[G_ReB] end
3130                                       >Sith[Ii2]; Fin[Mi2] end
3131                             >Sith[Ti]; Fin[Ti] end
3132                     >Sith[T]; Fin[T] end
3133             >structure end
3134             >Konw[Conc(C_ReB,G_ReB)]
3135              >E[Ч,e[Ч](C_ReB)]
3136              >E[Ч,e[Ч](G_ReB)]
3137             >Konw[Conc(C_ReB,G_ReB)]end
3138             >Konw[Causality_ReB]
3139              >of Sith[Subject_ReB], Sith[Subject_ReB,d_ReB(Subject_ReB)],
3140              >Sith[Subject_ReB,d_ReB(Subject_ReB),c_ReB(d_ReB(Subject_ReB))]
3141              >Sith[Object_ReB], Sith[Object_ReB,d_ReB(Object_ReB)],
3142              >Sith[Object_ReB,d_ReB(Object_ReB),c_ReB(d_ReB(Object_ReB))]
3143             >Konw[Causality_ReB]end
3144             >Konp[Obs5A_ReB_4]
3145              >of Sith[Obs5A_ReB_4], Fin[Obs]
3146              >E[Ц,e[Ц](5A_ReB_4)], E[Ц,e[Ц](R5)]
3147             >Konp[Obs5A_ReB_4]end
3148             >Konr[Descript((T,G_ReB)),R5)]
3149              >E[Ч,e[Ч](T)], E[Ч,e[Ч](G_ReB)]
3150              >E[Ц,e[Ц](R5)]
3151             >Konr[Descript((T,G_ReB)),R5)]end
3152             >Konr[Prior,Sits[R5]]
3153              >Sith[To]
3154              >Sith[Ti]
3155             >Konr[Prior,Sits[R5]]end
3156             >Konr[Prior,Sits[R5]]
3157              >Fin[Mo,x(Mo,p)]
3158              >Fin[Mo,x(Mo,p)+1]
3159             >Konr[Prior,Sits[R5]]end
3160             >Konr[Prior,Sits[R5]]
3161              >Fin[Mi,y(Mi,p)]
3162              >Fin[Mi,y(Mi,p)+1]
3163             >Konr[Prior,Sits[R5]]end
3164           >Sits[5A_AdB_4]end
3165     >Konf[Ex_ReB]
3166      >of Sits[5A_ReB_4]
3167      >Konr[Descript(G_ReB,R5)]
3168       >E[Ч,e[Ч](G_ReB)]
3169       >E[Ц,e[Ц](R5)]
3170      >Konr[Descript(G_ReB,R5)]end
3171      >if (Konr[Descript(G_ReB,R5)]=omnivial):
3172       >ExReBA=1
3173      >else:
```

```
3174        >ExReBA=0
3175    >endif
3176    >i_ReB=1
3177    >ExReBloop=1
3178    >while (ExReBloop=1):
3179     >if (ExReBA=1):
3180            >Sits[Modeli_ReB]
3181              >structure
3182                    >Sith[chi]; Fin[chi]
3183                            >Sith[Subject_ReB]
3184                                    >Sith[Subject_ReB,d_ReB(Subject_ReB)]
3185                                         >Sith[Subject_ReB,d_ReB(Subject_ReB),
3186                                         >c_ReB(d_ReB(Subject_ReB))]
3187                                    >Sith[Subject_ReB,d_ReB(Subject_ReB)] end
3188                            >Sith[Subject_ReB] end
3189                            >Sith[Object_ReB]
3190                                    >Sith[Object_ReB,d_ReB(Object_ReB)]
3191                                         >Sith[Object_ReB,d_ReB(Object_ReB),
3192                                         >c_ReB(d_ReB(Object_ReB))]
3193                                    >Sith[Object_ReB,d_ReB(Object_ReB)] end
3194                            >Sith[Object_ReB] end
3195                    >Sith[chi]; Fin[chi] end
3196              >structure end
3197              >Konw[NG_ReB]
3198               >of Sith[Subject_ReB], Sith[Subject_ReB,d_ReB(Subject_AdB)],
3199               >Sith[Subject_ReB,d_ReB(Subject_ReB),c_ReB(d_ReB(Subject_ReB))]
3200               >Sith[Object_ReB], Sith[Object_ReB,d_ReB(Object_ReB)],
3201               >Sith[Object_ReB,d_ReB(Object_ReB),c_ReB(d_ReB(Object_ReB))]
3202              >Konw[NG_ReB] end
3203            >Sits[Modeli_ReB] end
3204            >Sits[h,Ii_ReB]
3205              >structure
3206                    >Sith[chi]; Fin[chi]
3207                            >Sith[Subject_ReB]
3208                                    >Sith[Subject_ReB,d_ReB(Subject_ReB)]
3209                                         >Sith[Subject_ReB,d_ReB(Subject_ReB),
3210                                         >c_ReB(d_ReB(Subject_ReB))]
3211                                    >Sith[Subject_ReB,d_ReB(Subject_ReB)] end
3212                            >Sith[Subject_ReB] end
3213                    >Sith[chi]; Fin[chi] end
3214              >structure end
3215            >Sits[h,Ii_ReB] end
3216      >Class(Sith[chi],Fin[chi](Sits[h,Ii_ReB]))=
3217      >Konf[M](E[Ц,e[Ц](Modeli_ReB)],Konw[NG_ReB])
3218      >Konr[Subject_ReB,chi]
3219        >Sith[Subject_ReB]
3220        >Sith[chi]
3221       >Konr[Subject_ReB,chi]end
3222      >Konr[Subject_ReB,chi]=Konf[M](Konw[NG_ReB])
3223      >Class(Konr[Subject_ReB,chi])=Ef
3224      >Sith[Ii_ReB] Sith[Ii_ReB,y_ReB(Ii_ReB)]
3225       >E[Ч^5,e[Ч^5](Ii_ReB)]=Class(Sith[chi],Fin[chi])
3226      >Fin[Mi_ReB], Fin[Mi_ReB,y_ReB(Mi_ReB)]=Konw[M](Sits[Modeli_ReB])
3227       >E[Ч^5,e[Ч^5](Mi_ReB)]=E[3,e[3](Modeli_ReB)]
3228      >q_ReB(Ii_ReB)=abs(Class(Sith[chi],Fin[chi]))
3229      >q_ReB(Mi_ReB)=abs(ALL(E[3,e[3](Model_ReB)]))
3230      >Sits[5A_ReB_5]
3231     >elif (ExReBA=0):
3232      >s_ReB[1:i_ReB]
3233      >w_ReB1(i_ReB)=w_ReB1
```

```
3234    >w_ReB2(i_ReB)=w_ReB2
3235    >Sith[i_ReB,Subject_ReB(i_ReB),d(Subject_ReB(i_ReB)),
3236    >c_ReB(d_ReB(Subject_ReB(i_ReB)))]=
3237    >Sith[Subject_ReB,d_ReB(Subject_ReB),c_ReB(d_ReB(Subject_ReB))]
3238    >Sith[i_ReB,Subject_ReB,d_ReB(Subject(i_ReB))]=
3239    >Sith[Subject_ReB,d_ReB(Subject_ReB)]
3240    >Sith[i_ReB,Subject_AdB(i_ReB)]=Sith[Subject_ReB]
3241    >Sith[i_ReB,Object_ReB(i_ReB),d(Object_ReB(i_AdB)),
3242    >c_ReB(d_ReB(Object_ReB(i_ReB)))]=
3243    >Sith[Object_ReB,d_ReB(Object_ReB),c_ReB(d_ReB(Object_ReB))]
3244    >Sith[i_ReB,Object_ReB,d_ReB(Object(i_ReB))]=Sith[Object_ReB,d_ReB(Object_ReB)]
3245    >Sith[i_ReB,Object_ReB(i_ReB)]=Sith[Object_ReB]
3246        >Sits[F(i_ReB)]
3247         >structure
3248                >Sith[Guess,s_ReB]
3249                    >Sith[s_ReB,Subject(s_ReB)]
3250                        >Sith[s_ReB,Subject_ReB(s_ReB),d_ReB(Subject(s_ReB))]
3251                            >Sith[s_ReB,Subject_ReB(s_ReB),
3252                            >d_ReB(Subject_ReB(s_ReB)),
3253                            >c_ReB(d_ReB(Subject_ReB(s_ReB)))]
3254                        >Sith[s_ReB,Subject_ReB(s_ReB),
3255                        >d_ReB(Subject_ReB(s_ReB))] end
3256                    >Sith[s_ReB,Subject(s_ReB)] end
3257                    >Sith[s_ReB,Object(s_ReB)]
3258                        >Sith[s_ReB,Object_ReB(s_ReB),d_ReB(Object_ReB(s_ReB))]
3259                            >Sith[s_ReB,Object_ReB(s_ReB),
3260                            >d_ReB(Object_ReB(s_ReB)),
3261                            >c_ReB(d_ReB(Object_ReB(s_ReB)))]
3262                        >Sith[s_ReB,Object_ReB(s_ReB),
3263                        >d_ReB(Object_ReB(s_ReB))] end
3264                    >Sith[s_ReB,Object_ReB(s_ReB)] end
3265                >Sith[Guess,s_ReB] end
3266         >structure end
3267         >Konw[Causality_ReB]
3268          >of Sith[s_ReB,Subject_ReB(s_ReB)], Sith[s_ReB,Subject_ReB(s_ReB),
3269          >d_ReB(Subject_ReB(s_ReB))],
3270          >Sith[Subject_ReB(s_ReB),d_ReB(Subject_ReB(s_ReB)),
3271          >c_ReB(d_ReB(Subject_ReB(s_ReB)))]
3272          >Sith[s_ReB,Object_ReB(s_ReB)],
3273          >Sith[s_ReB,Object_ReB(s_ReB),d_ReB(Object_ReB(s_ReB))],
3274          >Sith[s_ReB,Object_ReB(s_ReB),d_ReB(Object_ReB(s_ReB)),
3275          >c_ReB(d_ReB(Object_ReB(s_ReB)))]
3276         >Konw[Causality_ReB]end
3277        >Sits[F(i_ReB)]end
3278    >w_ReB1,w_ReB2=groups(m_ReB)
3279     >w_ReB1\=w_ReB1(s_ReB)
3280     >w_ReB2\=w_ReB2(s_ReB)
3281    >Konr[w_ReB1,w_ReB2]
3282      >Sith[C,w_ReB1]
3283      >Sith[C,w_ReB2]
3284     >Konr[w_ReB1,w_ReB2]end
3285     >Sits[G]=Konf[int](Konr[w_ReB1,w_ReB2])
3286     >E[Ч,e[Ч](Subject_ReB)]=Sith[C,w_ReB1]
3287      >E[Ч,e[Ч](Object_ReB)]=Sith[C,w_ReB2]
3288    >Konw[Causality_ReB]=Class(Konr[w_ReB1,w_ReB2])
3289    >Sits[G_ReB]
3290     >E[Ц,e[Ц](G_ReB)]\=E[Ц,e[Ц](F(i_ReB))]
3291    >if (abs(Sits[G_ReB])\=0):
3292     >ExReBloop=1
3293    >elif (abs(Sits[G_ReB])=0):
```

```
3294        >ExReBloop=0
3295        >proceed
3296       >endif
3297          >Sits[G_ReB]
3298           >structure
3299               >Sith[Subject_ReB]
3300                     >Sith[Subject_ReB,d_ReB(Subject_ReB)]
3301                           >Sith[Subject_ReB,d_ReB(Subject_ReB),
3302                            >c_ReB(d_ReB(Subject_ReB))]
3303                     >Sith[Subject_ReB,d_ReB(Subject_ReB)] end
3304              >Sith[Subject_ReB] end
3305              >Sith[Object_ReB]
3306                     >Sith[Object_ReB,d_ReB(Object_ReB)]
3307                           >Sith[Object_ReB,d_ReB(Object_ReB),
3308                            >c_ReB(d_ReB(Object_ReB))]
3309                     >Sith[Object_ReB,d_ReB(Object_ReB)] end
3310              >Sith[Object_ReB] end
3311           >structure end
3312           >Konw[Causality_ReB]
3313            >of Sith[Subject_ReB], Sith[Subject_ReB,d_ReB(Subject_ReB)],
3314            >Sith[Subject_ReB,d_ReB(Subject_ReB),c_ReB(d_ReB(Subject_ReB))]
3315            >Sith[Object_ReB], Sith[Object_ReB,d_ReB(Object_ReB)],
3316            >Sith[Object_ReB,d_ReB(Object_ReB),c_ReB(d_ReB(Object_ReB))]
3317           >Konw[Causality_ReB]end
3318          >Sits[G_ReB] end
3319     >Sith[G_ReB]=Konw[M](Sits[G_ReB])
3320      >E[Ч,e[Ч](G_ReB)]=E[3,e[3](G_ReB)]
3321     >Kont[ReB(i_ReB)]
3322     >i_ReB=i_ReB+1
3323    >endif
3324    >else
3325    >Sits[5A_ReB_1]
3326    >Sits[5A_ReB]=Sits[5A_ReB_1]
3327     >E[Ц,e[Ц](5A_ReB)]=E[Ц,e[Ц](5A_ReB_1)]
3328    >endloop
3329   >Konf[Ex_ReB]end
3330          >Sits[5A_ReB_5]
3331           >structure
3332               >Sits[R5]
3333                >E[Ц,e[Ц](R5)]
3334               >Sith[Obs5A_ReB_5]; Fin[Obs]
3335               >Sith[T]; Fin[T]
3336                   >Sith[To]; Fin[To]
3337                       >Sith[Io,x(Io)]; Fin[Mo,x(Mo)]
3338                           >Sith[Io,x(Io),x2(x(Io))];
3339                           >Fin[Mo,x(Mo),x2(x(Mo))]
3340                       >Sith[Io,x(Io)]; Fin[Mo,x(Mo)] end
3341                   >Sith[To]; Fin[To] end
3342                   >Sith[Ti]; Fin[Ti]
3343                       >Sith[Ii1]; Fin[Mi1]
3344                           >Sith[Ii,y(Ii)], Fin[Mi,y(Mi)]
3345                               >Sith[Ii,y(Ii),y2(y(Ii))];
3346                               >Fin[Mi,y(Mi),y2(y(Mi))]
3347                           >Sith[Ii,y(Ii)], Fin[Mi,y(Mi)] end
3348                       >Sith[Ii2]
3349                           >Sith[Ii_ReB]; Fin[Mi_ReB]
3350                               >Sith[Ii_ReB,y_ReB(Ii_ReB)];
3351                               >Fin[Mi_ReB,y_ReB(Mi_ReB)]
3352                           >Sith[Ii_ReB]; Fin[Mi_ReB] end
3353                       >Sith[Ii2] end
```

```
3354                                          >Sith[Ii1]; Fin[Mi1] end
3355                                 >Sith[Ti]; Fin[Ti] end
3356                        >Sith[T]; Fin[T] end
3357                 >structure end
3358                 >Konp[Obs5A_ReB_5]
3359                  >of Sith[Obs5A_ReB_5], Fin[Obs]
3360                  >E[Ц,е[Ц](5A_AdB_5)], E[Ц,е[Ц](R5)]
3361                 >Konp[Obs5A_ReB_5]end
3362                 >Konr[Descript(T,R5)]
3363                  >E[Ч,е[Ч](T)]
3364                  >E[Ц,е[Ц](R5)]
3365                 >Konr[Descript(T,R5)]end
3366                 >Konr[Prior,Sits[R5]]
3367                  >Sith[To]
3368                  >Sith[Ti]
3369                 >Konr[Prior,Sits[R5]]end
3370                 >Konr[Prior,Sits[R5]]
3371                  >Fin[Mo,x(Mo,p)]
3372                  >Fin[Mo,x(Mo,p)+1]
3373                 >Konr[Prior,Sits[R5]]end
3374                 >Konr[Prior,Sits[R5]]
3375                  >Fin[Mi,y(Mi,p)]
3376                  >Fin[Mi,y(Mi,p)+1]
3377                 >Konr[Prior,Sits[R5]]end
3378                >Sits[5A_ReB_5]end
3379        >Konw[5A_ReB_6]
3380         >of Sits[5A_AdB_5]
3381        >Fin[Mi,y(Mi),y2(y(Mi))]=Fin[Mi,y(Mi),y2(y(Mi))]+Fin[Mi_ReB,y_ReB(Mi_ReB)]
3382        >Fin[Mi,y(Mi)]=Fin[Mi,y(Mi)]+Fin[Mi_ReB]
3383        >q(Mi)=q(Mi)-1
3384        >y(Mi)[1:q(Mi)]
3385        >y(Mi,p)[1:q(Mi)-1]
3386        >Sits[5A_ReB_6]
3387        >Konw[5A_ReB_6] end
3388                 >Sits[5A_ReB_6]
3389                  >structure
3390                      >Sits[R5]
3391                       >E[Ц,е[Ц](R5)]
3392                      >Sith[Obs5A_ReB_6]; Fin[Obs]
3393                      >Sith[T]; Fin[T]
3394                          >Sith[To]; Fin[To]
3395                              >Sith[Io,x(Io)]; Fin[Mo,x(Mo)]
3396                                  >Sith[Io,x(Io),x2(x(Io))];
3397                                  >Fin[Mo,x(Mo),x2(x(Mo))]
3398                              >Sith[Io,x(Io)]; Fin[Mo,x(Mo)] end
3399                          >Sith[To]; Fin[To] end
3400                          >Sith[Ti]; Fin[Ti]
3401                              >Sith[Ii,y(Ii)], Fin[Mi,y(Mi)]
3402                                  >Sith[Ii,y(Ii),y2(y(Ii))];
3403                                  >Fin[Mi,y(Mi),y2(y(Mi))]
3404                              >Sith[Ii,y(Ii)], Fin[Mi,y(Mi)] end
3405                          >Sith[Ti]; Fin[Ti] end
3406                      >Sith[T]; Fin[T] end
3407                 >structure end
3408                 >Konp[Obs5A_ReB_6]
3409                  >of Sith[Obs5A_ReB_6], Fin[Obs]
3410                  >E[Ц,е[Ц](5A_AdB_6)], E[Ц,е[Ц](R5)]
3411                 >Konp[Obs5A_ReB_6]end
3412                 >Konr[Descript(T,R5)]
3413                  >E[Ч,е[Ч](T)]
```

```
3414                     >E[Ц,e[Ц](R5)]
3415                  >Konr[Descript(T,R5)]end
3416                  >Konr[Prior,Sits[R5]]
3417                   >Sith[To]
3418                   >Sith[Ti]
3419                  >Konr[Prior,Sits[R5]]end
3420                  >Konr[Prior,Sits[R5]]
3421                   >Fin[Mo,x(Mo,p)]
3422                   >Fin[Mo,x(Mo,p)+1]
3423                  >Konr[Prior,Sits[R5]]end
3424                  >Konr[Prior,Sits[R5]]
3425                   >Fin[Mi,y(Mi,p)]
3426                   >Fin[Mi,y(Mi,p)+1]
3427                  >Konr[Prior,Sits[R5]]end
3428              >Sits[5A_ReB_6]end
3429        >Sits[5A_RcB]=Sits[5A_RcB_6]
3430         >E[Ц,e[Ц](5A_ReB)]=E[Ц,e[Ц](5A_ReB_6)]
3431     >Kont[ReBelt]end
3432     >Kont[ReB(i_ReB)]
3433              >Sits[5A_ReB_4(i_ReB)]
3434               >structure
3435                   >Sits[R5]
3436                    >E[Ц,e[Ц](R5)]
3437                   >Sith[Obs5A_AdB-4(i_ReB)]; Fin[Obs]
3438                   >Sith[T]; Fin[T]
3439                        >Sith[To]; Fin[To]
3440                            >Sith[Io,x(Io)]; Fin[Mo,x(Mo)]
3441                                 >Sith[Io,x(Io),x2(x(Io))];
3442                                   >Fin[Mo,x(Mo),x2(x(Mo))]
3443                            >Sith[Io,x(Io)]; Fin[Mo,x(Mo)] end
3444                        >Sith[To]; Fin[To] end
3445                        >Sith[Ti]; Fin[Ti]
3446                            >Sith[Ii1]; Fin[Mi1]
3447                                 >Sith[Ii,y(Ii)], Fin[Mi,y(Mi)]
3448                                      >Sith[Ii,y(Ii),y2(y(Ii))];
3449                                        >Fin[Mi,y(Mi),y2(y(Mi))]
3450                                 >Sith[Ii,y(Ii)], Fin[Mi,y(Mi)] end
3451                            >Sith[Ii1]; Fin[Mi1] end
3452                            >Sith[Ii2]; Fin[Mi2]
3453                                 >Sith[C_ReB]
3454                                     >Sith[C_ReB,w_ReB]
3455                                 >Sith[C_ReB] end
3456                                 >Sith[G_ReB]
3457                                     >Sith[Subject_ReB]
3458                                         >Sith[Subject_ReB
3459                                         >d_ReB(Subject_ReB)]
3460                                             >Sith[Subject_ReB,
3461                                             >d_ReB(Subject_ReB),
3462                                             >c_ReB(d_ReB(Subject_ReB))]
3463                                         >Sith[Subject_ReB,
3464                                         >d_ReB(Subject_ReB)] end
3465                                     >Sith[Subject_ReB] end
3466                                     >Sith[Object_ReB]
3467                                         >Sith[Object_ReB,
3468                                         >d_ReB(Object_ReB)]
3469                                             >Sith[Object_ReB,
3470                                             >d_ReB(Object_ReB),
3471                                             >c_ReB(d_ReB(Object_ReB))]
3472                                         >Sith[Object_ReB,
3473                                         >d_ReB(Object_ReB)] end
```

```
3474                                              >Sith[Object_ReB] end
3475                                         >Sith[G_ReB] end
3476                                >Sith[Ii2]; Fin[Mi2] end
3477                          >Sith[Ti]; Fin[Ti] end
3478                    >Sith[T]; Fin[T] end
3479              >structure end
3480              >Konw[Conc(C_ReB,G_ReB)]
3481               >E[Ч,e[Ч](C_ReB)]
3482               >E[Ч,e[Ч](G_ReB)]
3483              >Konw[Conc(C_ReB,G_ReB)]end
3484              >Konw[Causality_ReB]
3485               >of Sith[Subject_ReB], Sith[Subject_ReB,d_ReB(Subject_ReB)],
3486               >Sith[Subject_ReB,d_ReB(Subject_ReB),c_ReB(d_ReB(Subject_ReB))]
3487               >Sith[Object_ReB], Sith[Object_ReB,d_ReB(Object_ReB)],
3488               >Sith[Object_ReB,d_ReB(Object_ReB),c_ReB(d_ReB(Object_ReB))]
3489              >Konw[Causality_ReB]end
3490              >Konp[Obs5A_ReB_4]
3491               >of Sith[Obs5A_ReB_4(i_ReB)], Fin[Obs]
3492               >E[Ц,e[Ц](5A_ReB_4)(i_ReB)], E[Ц,e[Ц](R5)]
3493              >Konp[Obs5A_ReB_4(i_ReB)]end
3494              >Konr[Descript((T,(G_ReB)),R5)]
3495               >E[Ч,e[Ч](T)], E[Ч,e[Ч](G_ReB)]
3496               >E[Ц,e[Ц](R5)]
3497              >Konr[Descript((T,(G_ReB)),R5)]end
3498              >Konr[Prior,Sits[R5]]
3499               >Sith[To]
3500               >Sith[Ti]
3501              >Konr[Prior,Sits[R5]]end
3502              >Konr[Prior,Sits[R5]]
3503               >Fin[Mo,x(Mo,p)]
3504               >Fin[Mo,x(Mo,p)+1]
3505              >Konr[Prior,Sits[R5]]end
3506              >Konr[Prior,Sits[R5]]
3507               >Fin[Mi,y(Mi,p)]
3508               >Fin[Mi,y(Mi,p)+1]
3509              >Konr[Prior,Sits[R5]]end
3510             >Sits[5A_ReB_4]end
3511     >Kont[ReB(i_ReB)]end
3512     >Kont[NiCore]
3513      >Sits[5A_NiC_1]=Sits[5A(1[0])]
3514       >E[Ц,e[Ц](5A_NiC_1)]=E[Ц,e[Ц](5A(1[0]))]
3515           >Sits[5A_NiC_1]
3516             >structure
3517                 >Sits[R5]
3518                  >E[Ц,e[Ц](R5)]
3519                 >Sith[Obs5A_NiC_1]; Fin[Obs]
3520                 >Sith[T]; Fin[T]
3521                     >Sith[Io,x(Io)]; Fin[Mo,x(Mo)]
3522                         >Sith[Io,x(Io),x2(x(Io))]; Fin[Mo,x(Mo),x2(x(Mo))]
3523                     >Sith[Io,x(Io)]; Fin[Mo,x(Mo)] end
3524                 >Sith[T]; Fin[T] end
3525             >structure end
3526             >Konp[Obs5A_NiC_1]
3527              >of Sith[Obs5A_NiC_1], Fin[Obs]
3528              >E[Ц,e[Ц](5A_NiC_1)], E[Ц,e[Ц](R5)]
3529             >Konp[Obs5A_NiC_1]end
3530             >Konr[Descript(T,R5)]
3531              >E[Ч,e[Ч](T)]
3532              >E[Ц,e[Ц](R5)]
3533             >Konr[Descript(T,R5)]end
```

```
3534          >Konr[Prior,Sits[R5]]
3535           >Fin[Mo,x(Mo,p)]
3536           >Fin[Mo,x(Mo,p)+1]
3537           >Konr[Prior,Sits[R5]]end
3538         >Sits[5A_NiC_1]end
3539     >Konw[4NiBelt_2]
3540      >of Sits[5A_NiB_1]
3541      >Sits[5A_NiB_2]
3542       >y(Mo)[1:p(Mo)-1]
3543       >y(Mo,p)[1:p(Mo)-2]
3544       >Fin[Mo,x(Mo)]
3545        >\Fin[Mo,x(Mo,p)max+1]
3546       >p(Mo)=p(Mo)-1
3547       >x(Mo)[1:p(Mo)]
3548       >x(Mo,p)[1:p(Mo)-1]
3549     >Konw[4NiBelt_2]end
3550           >Sits[5A_NiC_2]
3551            >structure
3552               >Sits[R5]
3553                >E[Ц,e[Ц](R5)]
3554               >Sith[Obs5A_NiC_2]; Fin[Obs]
3555               >Sith[T]; Fin[T]
3556                  >Sith[Io,x(Io)]; Fin[Mo,x(Mo)]
3557                     >Sith[Io,x(Io),x2(x(Io))]; Fin[Mo,x(Mo),x2(x(Mo))]
3558                  >Sith[Io,x(Io)]; Fin[Mo,x(Mo)] end
3559               >Sith[T]; Fin[T] end
3560            >structure end
3561            >Konp[Obs5A_NiC_2]
3562             >of Sith[Obs5A_NiC_2], Fin[Obs]
3563             >E[Ц,e[Ц](5A_NiC_2)], E[Ц,e[Ц](R5)]
3564            >Konp[Obs5A_NiC_2]end
3565            >Konr[Descript(T,R5)]
3566             >E[Ч,e[Ч](T)]
3567             >E[Ц,e[Ц](R5)]
3568            >Konr[Descript(T,R5)]end
3569            >Konr[Prior,Sits[R5]]
3570             >Fin[Mo,x(Mo,p)]
3571             >Fin[Mo,x(Mo,p)+1]
3572            >Konr[Prior,Sits[R5]]end
3573          >Sits[5A_NiC_2]end
3574      >Sits[5A_NiC]=Sits[5A_Ni_2]
3575       >E[Ц,e[Ц](5A_NiC)]=E[Ц,e[Ц](5A_NiC_2)]
3576     >Kont[NiCore]end
3577     >Kont[AdCore]
3578      >Sits[5A_AdC_1]=Sits[5A(1[0])]
3579       >E[Ц,e[Ц](5A_AdC_1)]=E[Ц,e[Ц](5A(1[0]))]
3580           >Sits[5A_AdC_1]
3581            >structure
3582               >Sits[R5]
3583                >E[Ц,e[Ц](R5)]
3584               >Sith[Obs5A_AdC_1]; Fin[Obs]
3585               >Sith[T]; Fin[T]
3586                  >Sith[To]; Fin[To]
3587                     >Sith[Io,x(Io)]; Fin[Mo,x(Mo)]
3588                        >Sith[Io,x(Io),x2(x(Io))];
3589                        >Fin[Mo,x(Mo),x2(x(Mo))]
3590                     >Sith[Io,x(Io)]; Fin[Mo,x(Mo)] end
3591                  >Sith[To]; Fin[To] end
3592               >Sith[T]; Fin[T] end
3593            >structure end
```

```
3594        >Konp[Obs5A_AdC_1]
3595         >of Sith[Obs5A_AdC_1], Fin[Obs]
3596         >E[Ц,е[Ц](5A_AdC_1)], E[Ц,е[Ц](R5)]
3597        >Konp[Obs5A_AdC_1]end
3598        >Konr[Descript(T,R5)]
3599         >E[Ч,е[Ч](T)]
3600         >E[Ц,е[Ц](R5)]
3601        >Konr[Descript(T,R5)]end
3602        >Konr[Prior,Sits[R5]]
3603         >Fin[Mo,x(Mo,p)]
3604         >Fin[Mo,x(Mo,p)+1]
3605        >Konr[Prior,Sits[R5]]end
3606       >Sits[5A_AdC_1]end
3607    >Konw[4AdCore_2]
3608     >of Sits[5A_AdC_1]
3609     >Sits[5A_AdC_2]
3610      >x(Mo)[1:p(Mo)-1]
3611      >x(Mo,p)[1:p(Mo)-2]
3612      >Fin[Mo1]
3613      >\Fin[Mo,x(Mo,p)max+1]
3614      >Fin[Mo,AdCmax(Mo)]=Fin[Mo,x(Mo,p)max+1]
3615    >Konw[4AdCore_2]end
3616           >Sits[5A_AdC_2]
3617            >structure
3618               >Sits[R5]
3619                >E[Ц,е[Ц](R5)]
3620               >Sith[Obs5A_AdC_2]; Fin[Obs]
3621               >Sith[T]; Fin[T]
3622                   >Sith[To]; Fin[To]
3623                       >Sith[Io1]; Fin[Mo1]
3624                           >Sith[Io,x(Io)]; Fin[Mo,x(Mo)]
3625                               >Sith[Io,x(Io),x2(x(Io))];
3626                               >Fin[Mo,x(Mo),x2(x(Mo))]
3627                           >Sith[Io,x(Io)]; Fin[Mo,x(Mo)] end
3628                       >Sith[Io1]; Fin[Mo1] end
3629                       >Sith[Io,AdCmax(Io)]; Fin[Mo,AdCmax(Mo)]
3630                           >Sith[Io,AdCmax(Io),x2(AdCmax(Io))];
3631                           >Fin[Mo,AdCmax(Mo),x2(AdCmax(Mo))]
3632                       >Sith[Io,AdCmax(Io)]; Fin[Mo,AdCmax(Mo)] end
3633                   >Sith[To]; Fin[To] end
3634               >Sith[T]; Fin[T] end
3635            >structure end
3636           >Konp[Obs5A_AdC_2]
3637            >of Sith[Obs5A_AdC_2], Fin[Obs]
3638            >E[Ц,е[Ц](5A_AdC_2)], E[Ц,е[Ц](R5)]
3639           >Konp[Obs5A_AdC_2]end
3640           >Konr[Descript(T,R5)]
3641            >E[Ч,е[Ч](T)]
3642            >E[Ц,е[Ц](R5)]
3643           >Konr[Descript(T,R5)]end
3644           >Konr[Prior,Sits[R5]]
3645            >Fin[Mo,x(Mo,p)]
3646            >Fin[Mo,x(Mo,p)+1]
3647           >Konr[Prior,Sits[R5]]end
3648          >Sits[5A_AdC_2]end
3649    >Konw[4AdCore_3]
3650     >of Sits[5A_AdC_2]
3651     >Sith[C1_AdC,c_AdC]=mutual(Sith[R5,a_AdC,b_AdC(a_AdC)])
3652     >while (c_AdC=0):
3653       >\Sits[R5]
```

```
3654            >Sits[R5new]\=Sits[R5]
3655                >Sits[R5new]
3656                 >structure
3657                  >Sith[R5new]; Fin[R5new]
3658                          >Sith[R5new,a_AdCnew]; Fin[R5new,a_AdCnew]
3659                              >Sith[R5new,a_AdCnew,b_AdCnew(a_AdCnew)]
3660                          >Sith[R5new,a_AdCnew]; Fin[R5new,a_AdCnew] end
3661                  >Sith[R5new]; Fin[R5new] end
3662              >strucutre end
3663             >Sit[R5new] end
3664        >Sits[R5]=Sits[R5new]
3665         >E[Ц,e[Ц](R5)=E[Ц,e[Ц](R5new)
3666        >else:
3667              >Sits[C1_AdC]
3668               >structure
3669                 >Sith[C1_AdC]
3670                         >Sith[C1_AdC,c_AdC]
3671                  >Sith[C1_AdC] end
3672              >structure end
3673             >Sits[C1_AdC] end
3674        >Sith[d_AdC]=Classes(Sith[C1_AdC])
3675              >Sits[C2_AdC]
3676               >structure
3677                 >Sith[d_AdC]
3678                         >Sith[d_AdC,c_AdC(d_AdC)]
3679                  >Sith[d_AdC] end
3680              >structure end
3681             >Sits[C2_AdC] end
3682        >Sith[C_AdC], Fin[C_AdC]=Konw[M](Sits[C2_AdC])
3683         >E[Ч,e[Ч](C_AdC)]=E[3,e[3](C2_AdC)]
3684        >Sits[5A_AdC_3]
3685       >endloop
3686   >Konw[4AdCore_3]end
3687              >Sits[5A_AdC_3]
3688               >structure
3689                 >Sits[R5]
3690                  >E[Ц,e[Ц](R5)]
3691                 >Sith[Obs5A_AdC_3]; Fin[Obs]
3692                 >Sith[T]; Fin[T]
3693                     >Sith[To]; Fin[To]
3694                         >Sith[Io1]; Fin[Mo1]
3695                             >Sith[Io,x(Io)]; Fin[Mo,x(Mo)]
3696                                 >Sith[Io,x(Io),x2(x(Io))];
3697                                 >Fin[Mo,x(Mo),x2(x(Mo))]
3698                             >Sith[Io,x(Io)]; Fin[Mo,x(Mo)] end
3699                         >Sith[Io1]; Fin[Mo1] end
3700                         >Sith[Io2]; Fin[Mo2]
3701                             >Sith[Io,AdCmax(Io)]; Fin[Mo,AdCmax(Mo)]
3702                                 >Sith[Io,AdCmax(Io),x2(AdCmax(Io))];
3703                                 >Fin[Mo,AdCmax(Mo),x2(AdCmax(Mo))]
3704                             >Sith[Io,AdCmax(Io)]; Fin[Mo,AdCmax(Mo)] end
3705                             >Sith[C_AdC]
3706                                 >Sith[C_AdC,w_AdC]
3707                             >Sith[C_AdC] end
3708                         >Sith[Io2]; Fin[Mo2] end
3709                     >Sith[To]; Fin[To] end
3710                 >Sith[T]; Fin[T] end
3711              >structure end
3712             >Konp[Obs6A_AdC_3]
3713              >of Sith[Obs5A_AdC_3], Fin[Obs]
```

```
3714          >Е[Ц,е[Ц](5A_AdC_3)], Е[Ц,е[Ц](R5)]
3715          >Konp[Obs6A_AdC_3]end
3716          >Konr[Descript(T,R5)]
3717           >Е[Ч,е[Ч](T)]
3718           >Е[Ц,е[Ц](R5)]
3719          >Konr[Descript(T,R5)]end
3720          >Konr[R,C]
3721           >Е[Ц,е[Ц](R5)]
3722           >Е[Ц,е[Ц](Ex4oC_AdC)]
3723          >Konr[R,C]end
3724          >Konr[Prior,Sits[R5]]
3725           >Fin[Mo,x(Mo,p)]
3726           >Fin[Mo,x(Mo,p)+1]
3727          >Konr[Prior,Sits[R5]]end
3728         >Sits[5A_AdC_3]end
3729     >Konw[4AdCore_4]
3730      >of Sits[5A_AdC_3]
3731       >w_AdC1,w_AdC2=groups(w_AdC)
3732      >w_AdC1\=w_AdC2
3733      >Konr[w_AdC1,w_AdC2]
3734        >Sith[C,w_AdC1]
3735        >Sith[C,w_AdC2]
3736      >Konr[w_AdC1,w_AdC2]end
3737      >Sits[G_AdC]=Konf[int](Konr[w_AdC1,w_AdC2])
3738      >Е[Ч,е[Ч](Subject_AdC)]=Sith[C,w_AdC1]
3739       >Е[Ч,е[Ч](Object_AdC)]=Sith[C,w_AdC2]
3740      >Konw[Causality_AdC]=Class(Konr[w_AdC1,w_AdC2])
3741           >Sits[G_AdC]
3742            >structure
3743                >Sith[G_AdC]
3744                  >Sith[Subject_AdC]
3745                      >Sith[Subject_AdC,d_AdC(Subject_AdC)]
3746                         >Sith[Subject_AdC,d_AdC(Subject_AdC),
3747                          >c_AdC(d_AdC(Subject_AdC))]
3748                      >Sith[Subject_AdC,d_AdC(Subject_AdC)] end
3749                  >Sith[Subject_AdC] end
3750                  >Sith[Object_AdC]
3751                      >Sith[Object_AdC,d_AdC(Object_AdC)]
3752                         >Sith[Object_AdC,d_AdC(Object_AdC),
3753                          >c_AdC(d_AdC(Object_AdC))]
3754                      >Sith[Object_AdC,d_AdC(Object_AdC)] end
3755                  >Sith[Object_AdC] end
3756              >Sith[G_AdC] end
3757          >structure end
3758          >Konw[Causality_AdC]
3759           >of Sith[Subject_AdC], Sith[Subject_AdC,d_AdC(Subject_AdC)],
3760           >Sith[Subject_AdC,d_AdC(Subject_AdC),c_AdC(d_AdC(Subject_AdC))]
3761           >Sith[Object_AdC], Sith[Object_AdC,d_AdC(Object_AdC)],
3762           >Sith[Object_AdC,d_AdC(Object_AdC),c_AdC(d_AdC(Object_AdC))]
3763          >Konw[Causality_AdC]end
3764         >Sits[G_AdC] end
3765      >Sith[C_AdC]=Konw[M](Sits[C_AdC])
3766       >Е[Ч,е[Ч](G_AdC)]=E[3,e[3](G_AdC)]
3767      >Sits[5A_AdC_4]
3768     >Konw[4AdCore_4] end
3769          >Sits[5A_AdC_4]
3770           >structure
3771               >Sits[R5]
3772                >Е[Ц,е[Ц](R5)]
3773               >Sith[Obs5A_AdC-4]; Fin[Obs]
```

```
3774          >Sith[T]; Fin[T]
3775            >Sith[To]; Fin[To]
3776              >Sith[Io1]; Fin[Mo1]
3777                >Sith[Io,x(Io)]; Fin[Mo,x(Mo)]
3778                  >Sith[Io,x(Io),x2(x(Io))];
3779                  >Fin[Mo,x(Mo),x2(x(Mo))]
3780                >Sith[Io,x(Io)]; Fin[Mo,x(Mo)] end
3781              >Sith[Io1]; Fin[Mo1] end
3782              >Sith[Io2]; Fin[Mo2]
3783                >Sith[Ii,AdCmax(Ii)]; Fin[Mi,AdCmax(Mi)]
3784                  >Sith[Ii,AdCmax(Ii),y2(AdCmax(Ii))];
3785                  >Fin[Mi,AdCmax(Mi),y2(AdCmax(Mi))]
3786                >Sith[Ii,AdCmax(Ii)]; Fin[Mi,AdCmax(Mi)] end
3787              >Sith[C_AdC]
3788                >Sith[C_AdC,w_AdC]
3789              >Sith[C_AdC] end
3790              >Sith[G_AdC]
3791                >Sith[Subject_AdC]
3792                  >Sith[Subject_AdC,
3793                  >d_AdC(Subject_AdC)]
3794                    >Sith[Subject_AdC,
3795                    >d_AdC(Subject_AdC),
3796                    >c_AdC(d_AdC(Subject_AdC))]
3797                  >Sith[Subject_AdC,
3798                  >d_AdC(Subject_AdC)] end
3799                >Sith[Subject_AdC] end
3800                >Sith[Object_AdC]
3801                  >Sith[Object_AdC,
3802                  >d_AdC(Object_AdC)]
3803                    >Sith[Object_AdC,
3804                    >d_AdC(Object_AdC)
3805                    >c_AdC(d_AdC(Object_AdC))]
3806                  >Sith[Object_AdC,
3807                  >d_AdC(Object_AdC)] end
3808                >Sith[Object_AdC] end
3809              >Sith[G_AdC] end
3810            >Sith[Io2]; Fin[Mo2] end
3811          >Sith[To]; Fin[To] end
3812        >Sith[T]; Fin[T] end
3813      >structure end
3814      >Konw[Conc(C_AdC,G_AdC)]
3815        >E[Ч,e[Ч](C_AdC)]
3816        >E[Ч,e[Ч](G_AdC)]
3817      >Konw[Conc(C_AdC,G_AdC)]end
3818      >Konw[Causality_AdC]
3819        >of Sith[Subject_AdC], Sith[Subject_AdC,d_AdC(Subject_AdC)],
3820        >Sith[Subject_AdC,d_AdC(Subject_AdC),c_AdC(d_AdC(Subject_AdC))]
3821        >Sith[Object_AdC], Sith[Object_AdC,d_AdC(Object_AdC)],
3822        >Sith[Object_AdC,d_AdC(Object_AdC),c_AdC(d_AdC(Object_AdC))]
3823      >Konw[Causality_AdC]end
3824      >Konp[Obs5A_AdC_4]
3825        >of Sith[Obs5A_AdC_4], Fin[Obs]
3826        >E[Ц,e[Ц](5A_AdC_4)], E[Ц,e[Ц](R5)]
3827      >Konp[Obs5A_AdC_4]end
3828      >Konr[Descript((T,(G_AdC)),R5)]
3829        >E[Ч,e[Ч](T)], E[Ч,e[Ч](G_AdC)]
3830        >E[Ц,e[Ц](R5)]
3831      >Konr[Descript((T,(G_AdC)),R5)]end
3832      >Konr[Prior,Sits[R5]]
3833        >Fin[Mo,x(Mo,p)]
```

```
3834                        >Fin[Mo,x(Mo,p)+1]
3835                         >Konr[Prior,Sits[R5]]end
3836                       >Sits[5A_AdC_4]end
3837              >Konf[Ex_AdC]
3838               >of Sits[5A_AdC_4]
3839               >Konr[Descript(G_AdC,R5)]
3840                >E[Ч,e[Ч](G_AdC)]
3841                >E[Ц,e[Ц](R5)]
3842               >Konr[Descript(G_AdC,R5)]end
3843               >if (Konr[Descript(G_AdC,R5)]=omnivial):
3844                >ExAdCA=1
3845               >else:
3846                >ExAdCA=0
3847               >endif
3848              >i_AdC=1
3849              >ExAdCloop=1
3850              >while (ExAdCloop=1):
3851               >if (ExAdCA=1):
3852                    >Sits[Modelo_AdC]
3853                     >structure
3854                        >Sith[chi]; Fin[chi]
3855                            >Sith[Subject_AdC]
3856                                >Sith[Subject_AdC,d_AdC(Subject_AdC)]
3857                                    >Sith[Subject_AdC,d_AdC(Subject_AdC),
3858                                    >c_AdC(d_AdC(Subject_AdC))]
3859                                >Sith[Subject_AdC,d_AdC(Subject_AdC)] end
3860                            >Sith[Subject_AdC] end
3861                            >Sith[Object_AdC]
3862                                >Sith[Object_AdC,d_AdC(Object_AdC)]
3863                                    >Sith[Object_AdC,d_AdC(Object_AdC),
3864                                    >c_AdC(d_AdC(Object_AdC))]
3865                                >Sith[Object_AdC,d_AdC(Object_AdC)] end
3866                            >Sith[Object_AdC] end
3867                        >Sith[chi]; Fin[chi] end
3868                 >structure end
3869                 >Konw[NG_AdC]
3870                  >of Sith[Subject_AdC], Sith[Subject_AdC,d_AdC(Subject_AdC)],
3871                  >Sith[Subject_AdC,d_AdC(Subject_AdC),c_AdC(d_AdC(Subject_AdC))]
3872                  >Sith[Object_AdC], Sith[Object_AdC,d_AdC(Object_AdC)],
3873                  >Sith[Object_AdC,d_AdC(Object_AdC),c_AdC(d_AdC(Object_AdC))]
3874                 >Konw[NG_AdC] end
3875                >Sits[Modelo_AdC]] end
3876                >Sits[h,Io_AdC]]
3877                 >structure
3878                        >Sith[chi]; Fin[chi]
3879                            >Sith[Subject_AdC]
3880                                >Sith[Subject_AdC,d_AdC(Subject_AdC)]
3881                                    >Sith[Subject_AdC,d_AdC(Subject_AdC),
3882                                    >c_AdC(d_AdC(Subject_AdC))]
3883                                >Sith[Subject_AdC,d_AdC(Subject_AdC)] end
3884                            >Sith[Subject_AdC] end
3885                        >Sith[chi]; Fin[chi] end
3886                 >structure end
3887                >Sits[h,Io_AdC] end
3888              >Class(Sith[chi],Fin[chi](Sits[h,Io,_AdC]))=
3889              >Konf[M](E[Ц,e[Ц](Modelo_AdC)],Konw[NG_AdC])
3890              >Konr[Subject_AdC,chi]
3891                >Sith[Subject_AdC]
3892                >Sith[chi]
3893               >Konr[Subject_AdC,chi]end
```

```
3894        >Konr[Subject_AdC,chi]=Konf[M](Konw[NG_AdC])
3895        >Class(Konr[Subject_AdC,chi])=Ef
3896        >Sith[Io_AdC] Sith[Io_AdC,x_AdC(Io_AdC)]
3897         >E[4^5,e[4^5](Io_AdC)]=Class(Sith[chi],Fin[chi])
3898        >Fin[Mo_AdC], Fin[Mo_AdC,x_AdC(Mo_AdC)]=Konw[M](Sits[Modelo_AdC])
3899         >E[4^5,e[4^5](Mo_AdC)]=E[3,e[3](Modelo_AdC)]
3900        >q_AdC(Io_AdC)=abs(Class(Sith[chi],Fin[chi]))
3901        >q_AdC(Mo_AdC)=abs(ALL(E[3,e[3](Modelo_AdC)]))
3902        >Sits[5A_AdC_5]
3903      >elif (ExAdCA=0):
3904      >s_AdC[1:i_AdC]
3905      >w_AdC1(i_AdC)=w_AdC1
3906      >w_AdC2(i_AdC)=w_AdC2
3907      >Sith[i_AdC,Subject_AdC(i_AdC),d(Subject_AdC(i_AdC)),
3908      >c_AdC(d_AdC(Subject_AdC(i_AdC)))]=
3909      >Sith[Subject_AdC,d_AdC(Subject_AdC),c_AdC(d_AdC(Subject_AdC))]
3910      >Sith[i_AdC,Subject_AdC,d_AdC(Subject(i_AdC))]=
3911      >Sith[Subject_AdC,d_AdC(Subject_AdC)]
3912      >Sith[i_AdC,Subject_AdC(i_AdC)]=Sith[Subject_AdC]
3913      >Sith[i_AdC,Object_AdC(i_AdC),d(Object_AdC(i_AdC)),
3914      >c_AdC(d_AdC(Object_AdC(i_AdC)))]=
3915      >Sith[Object_AdC,d_AdC(Object_AdC),c_AdC(d_AdC(Object_AdC))]
3916      >Sith[i_AdC,Object_AdC,d_AdC(Object(i_AdC))]=Sith[Object_AdC,d_AdC(Object_AdC)]
3917      >Sith[i_AdC,Object_AdC(i_AdC)]=Sith[Object_AdC]
3918          >Sits[F(i_AdC)]
3919           >structure
3920               >Sith[Guess,s_AdC]
3921                   >Sith[s_AdB,Subject(s_AdC)]
3922                       >Sith[s_AdC,Subject_AdC(s_AdC),d_AdC(Subject(s_AdC))]
3923                           >Sith[s_AdC,Subject_AdC(s_AdC),
3924                           >d_AdC(Subject_AdC(s_AdC)),
3925                           >c_AdC(d_AdC(Subject_AdC(s_AdC)))]
3926                       >Sith[s_AdC,Subject_AdC(s_AdC),
3927                       >d_AdC(Subject_AdC(s_AdC))] end
3928                   >Sith[s_AdC,Subject(s_AdC)] end
3929                   >Sith[s_AdC,Object(s_AdC)]
3930                       >Sith[s_AdC,Object_AdC(s_AdC),d_AdC(Object_AdC(s_AdC))]
3931                           >Sith[s_AdC,Object_AdC(s_AdC),
3932                           >d_AdB(Object_AdC(s_AdC)),
3933                           >c_AdC(d_AdC(Object_AdC(s_AdC)))]
3934                       >Sith[s_AdC,Object_AdC(s_AdC),
3935                       >d_AdC(Object_AdC(s_AdC))] end
3936                   >Sith[s_AdC,Object_AdC(s_AdC)] end
3937               >Sith[Guess,s_AdC] end
3938          >structure end
3939          >Konw[Causality_AdC]
3940           >of Sith[s_AdC,Subject_AdC(s_AdC)], Sith[s_AdC,Subject_AdC(s_AdC),
3941           >d_AdC(Subject_AdC(s_AdC))],
3942           >Sith[Subject_AdC(s_AdC),d_AdC(Subject_AdC(s_AdC)),
3943           >c_AdC(d_AdC(Subject_AdC(s_AdC)))]
3944           >Sith[s_AdC,Object_AdC(s_AdC)], >Sith[s_AdC,Object_AdC(s_AdC),
3945           >d_AdC(Object_AdC(s_AdC))],Sith[s_AdC,Object_AdC(s_AdC),
3946           >d_AdC(Object_AdC(s_AdC)),c_AdC(d_AdC(Object_AdC(s_AdC)))]
3947           >Konw[Causality_AdC]end
3948          >Sits[F(i_AdC)]end
3949      >w_AdC1,w_AdC2=groups(m_AdC)
3950       >w_AdC1\=w_AdC1(s_AdC)
3951       >w_AdC2\=w_AdC2(s_AdC)
3952      >Konr[w_AdC1,w_AdC2]
3953       >Sith[C,w_AdC1]
```

```
3954        >Sith[C,w_AdC2]
3955       >Konr[w_AdC1,w_AdC2]end
3956       >Sits[G]=Konf[int](Konr[w_AdC1,w_AdC2])
3957       >E[Ч,e[Ч](Subject_AdC)]=Sith[C,w_AdC1]
3958        >E[Ч,e[Ч](Object_AdC)]=Sith[C,w_AdC2]
3959      >Konw[Causality_AdC]=Class(Konr[w_AdC1,w_AdC2])
3960      >Sits[G_AdC]
3961       >E[Ц,e[Ц](G_AdC)]\=E[Ц,e[Ц](F(i_AdC))]
3962      >if (abs(Sits[G_AdC])\=0):
3963       >ExAdCloop=1
3964      >elif (abs(Sits[G_AdC])=0):
3965       >ExAdCloop=0
3966       >proceed
3967      >endif
3968          >Sits[G_AdC]
3969           >structure
3970               >Sith[Subject_AdC]
3971                   >Sith[Subject_AdC,d_AdC(Subject_AdC)]
3972                       >Sith[Subject_AdC,d_AdC(Subject_AdC),
3973                       >c_AdC(d_AdC(Subject_AdC))]
3974                   >Sith[Subject_AdC,d_AdC(Subject_AdC)] end
3975               >Sith[Subject_AdC] end
3976               >Sith[Object_AdC]
3977                   >Sith[Object_AdC,d_AdC(Object_AdC)]
3978                       >Sith[Object_AdC,d_AdC(Object_AdC),
3979                       >c_AdC(d_AdC(Object_AdC))]
3980                   >Sith[Object_AdC,d_AdC(Object_AdC)] end
3981               >Sith[Object_AdC] end
3982           >structure end
3983           >Konw[Causality_AdC]
3984            >of Sith[Subject_AdC], Sith[Subject_AdC,d_AdC(Subject_AdC)],
3985            >Sith[Subject_AdC,d_AdC(Subject_AdC),c_AdC(d_AdC(Subject_AdC))]
3986            >Sith[Object_AdC], Sith[Object_AdC,d_AdC(Object_AdC)],
3987            >Sith[Object_AdC,d_AdC(Object_AdC),c_AdC(d_AdC(Object_AdC))]
3988           >Konw[Causality_AdC]end
3989          >Sits[G_AdC] end
3990      >Sith[C_AdC]=Konw[M](Sits[C_AdC])
3991       >E[Ч,e[Ч](G_AdC)]=E[3,e[3](G_AdC)]
3992      >Kont[AdC(i_AdC)]
3993      >i_AdC=i_AdC+1
3994     >endif
3995     >else
3996     >Sits[5A_AdC_1]
3997     >Sits[5A_AdC]=Sits[5A_AdC_1]
3998      >E[Ц,e[Ц](5A_AdC)]=E[Ц,e[Ц](5A_AdC_1)]
3999     >endloop
4000    >Konf[Ex_AdC]end
4001         >Sits[5A_AdC_5]
4002          >structure
4003              >Sits[R5]
4004               >E[Ц,e[Ц](R5)]
4005              >Sith[Obs5A_AdC_5]; Fin[Obs]
4006              >Sith[T]; Fin[T]
4007                  >Sith[Ti]; Fin[Ti]
4008                      >Sith[Io1]; Fin[Mo]
4009                          >Sith[Io,x(Io)]; Fin[Mo,x(Mo)]
4010                              >Sith[Io,x(Io),x2(x(Io))];
4011                              >Fin[Mo,x(Mo),x2(x(Mo))]
4012                          >Sith[Io,x(Io)]; Fin[Mo,x(Mo)] end
4013                      >Sith[Io2]; Fin[Mo2]
```

```
4014                                              >Sith[Io,AdCmax(Io)]; Fin[Mo,AdCmax(Mo)]
4015                                                    >Sith[Io,AdCmax(Io),
4016                                                    >x2(AdCmax(Io))];
4017                                                       >Fin[Mo,AdCmax(Mo),x2(AdCmax(Mo))]
4018                                              >Sith[Io,AdCmax(Io)];
4019                                              >Fin[Mo,AdCmax(Mo)] end
4020                                              >Sith[Io_AdC]; Fin[Mo_AdC]
4021                                                    >Sith[Io_AdC,x_AdC(Io_AdC)];
4022                                                       >Fin[Mo_AdC,x_AdC(Mo_AdC)]
4023                                              >Sith[Io_AdC]; Fin[Mo_AdC] end
4024                                      >Sith[Io2]; Fin[Mo2] end
4025                                >Sith[Io1]; Fin[Mo] end
4026                          >Sith[Ti]; Fin[Ti] end
4027                    >Sith[T]; Fin[T] end
4028              >structure end
4029              >Konp[Obs5A_AdC_5]
4030               >of Sith[Obs5A_AdC_5], Fin[Obs]
4031               >E[Ц,e[Ц](5A_AdC_5)], E[Ц,e[Ц](R5)]
4032              >Konp[Obs5A_AdC_5]end
4033              >Konr[Descript(T,R5)]
4034               >E[Ч,e[Ч](T)]
4035               >E[Ц,e[Ц](R5)]
4036              >Konr[Descript(T,R5)]end
4037              >Konr[Prior,Sits[R5]]
4038               >Fin[Mo,x(Mo,p)]
4039               >Fin[Mo,x(Mo,p)+1]
4040              >Konr[Prior,Sits[R5]]end
4041              >Sits[5A_AdC_5]end
4042        >Konw[5A_AdC_6]
4043         >of Sits[5A_AdB_5]
4044              >Sits[To2]
4045              >structure
4046                  >Sits[Io2,new]
4047                       >Sith[Io,AdCmax(Io),x2(AdCmax(Io))]; Fin[Mo,AdCmax(Mo),
4048                       >x2(AdCmax(Mo))]
4049                       >Sith[Io_AdC,x_AdC(Mo_AdC)]; Fin[Mo_AdC,x_AdC(Mo_AdC)]
4050                  >Sits[Io2,new] end
4051              >structure end
4052              >Sits[To2] end
4053        >Sith[Mo2new,x(Mo2new)]=Fin[Mo,AdCmax(Mo),x2(AdCmax(Mo))]+Fin[Mo_AdC,x_AdC(Mo_AdC)]
4054        >\Fin[Mi2]
4055        >Fin[Mo2]=Fin[Mo2new]
4056         >E[Ч,e[Ч](Mo2)]=E[Ч,e[Ч](Mo2new)]
4057        >Sits[5A_AdB_6]
4058        >Konw[5A_AdC_6] end
4059              >Sits[5A_AdC_6]
4060               >structure
4061                  >Sits[R5]
4062                   >E[Ц,e[Ц](R5)]
4063                  >Sith[Obs5A_AdC_6]; Fin[Obs]
4064                  >Sith[T]; Fin[T]
4065                       >Sith[Ti]; Fin[Ti]
4066                          >Sith[Io1]; Fin[Mo]
4067                             >Sith[Io,x(Io)]; Fin[Mo,x(Mo)]
4068                                >Sith[Io,x(Io),x2(x(Io))];
4069                                >Fin[Mo,x(Mo),x2(x(Mo))]
4070                             >Sith[Io,x(Io)]; Fin[Mo,x(Mo)] end
4071                          >Sith[Io2]; Fin[Mo2]
4072                             >Sith[Io2,x(Io2)]; Fin[Mo2,x(Mo2)]
4073                          >Sith[Io2]; Fin[Mo2] end
```

```
4074                                      >Sith[Io1]; Fin[Mo] end
4075                              >Sith[Ti]; Fin[Ti] end
4076                      >Sith[T]; Fin[T] end
4077              >structure end
4078              >Konp[Obs5A_AdC_6]
4079               >of Sith[Obs5A_AdC_5], Fin[Obs]
4080               >E[Ц,е[Ц](5A_AdC_6)], E[Ц,е[Ц](R5)]
4081              >Konp[Obs5A_AdC_6]end
4082              >Konr[Descript(T,R5)]
4083               >E[Ч,е[Ч](T)]
4084               >E[Ц,е[Ц](R5)]
4085              >Konr[Descript(T,R5)]end
4086              >Konr[Prior,Sits[R5]]
4087               >Fin[Mo,x(Mo,p)]
4088               >Fin[Mo,x(Mo,p)+1]
4089              >Konr[Prior,Sits[R5]]end
4090             >Sits[5A_AdC_6]end
4091      >Konw[5A_AdC_7]
4092       >of Sits[5A_AdC_6]
4093      >Fin[Mo,x(Mo),x2(x(Mo))]=Fin[Mo,x(Mo),x2(x(Mo))]+Fin[Mo2,y(Mo2)]
4094      >Fin[Mo,x(Mo)]=Fin[Mo,x(Mo)]+Fin[Mo2]
4095      >p(Mo)=p(Mo)-1
4096      >x(Mo)[1:p(Mo)]
4097      >x(Mo,p)[1:p(Mo)-1]
4098      >Sits[5A_AdC_7]
4099      >Konw[5A_AdC_7] end
4100             >Sits[5A_AdC_7]
4101               >structure
4102                  >Sits[R5]
4103                   >E[Ц,е[Ц](R5)]
4104                  >Sith[Obs5A_AdC_7]; Fin[Obs]
4105                  >Sith[T]; Fin[T]
4106                       >Sith[Ti]; Fin[Ti]
4107                            >Sith[Io,x(Io)]; Fin[Mo,x(Mo)]
4108                                 >Sith[Io,x(Io),x2(x(Io))];
4109                                 >Fin[Mo,x(Mo),x2(x(Mo))]
4110                            >Sith[Io,x(Io)]; Fin[Mo,x(Mo)] end
4111                       >Sith[Ti]; Fin[Ti] end
4112                  >Sith[T]; Fin[T] end
4113               >structure end
4114              >Konp[Obs5A_AdC_7]
4115               >of Sith[Obs5A_AdC_5], Fin[Obs]
4116               >E[Ц,е[Ц](5A_AdC_7)], E[Ц,е[Ц](R5)]
4117              >Konp[Obs5A_AdC_7]end
4118              >Konr[Descript(T,R5)]
4119               >E[Ч,е[Ч](T)]
4120               >E[Ц,е[Ц](R5)]
4121              >Konr[Descript(T,R5)]end
4122              >Konr[Prior(o)]
4123               >Fin[Mo,x(Mo,p)]
4124               >Fin[Mo,x(Mo,p)+1]
4125              >Konr[Prior(o)]end
4126             >Sits[5A_AdC_6]end
4127       >Sits[5A_AdC]=Sits[5A_AdC_7]
4128        >E[Ц,е[Ц](5A_AdC)]=E[Ц,е[Ц](5A_AdC_7)]
4129      >Kont[AdCore]end
4130      >Kont[AdC(i_AdC)]
4131             >Sits[5A_AdC_4(i_AdC)]
4132               >structure
4133                  >Sits[R5]
```

```
4134                          >E[Ц,e[Ц](R5)]
4135                      >Sith[Obs5A_AdC-4(i_AdC)]; Fin[Obs]
4136                      >Sith[T]; Fin[T]
4137                          >Sith[To]; Fin[To]
4138                              >Sith[Io1]; Fin[Mo1]
4139                                  >Sith[Io,x(Io)]; Fin[Mo,x(Mo)]
4140                                      >Sith[Io,x(Io),x2(x(Io))];
4141                                      >Fin[Mo,x(Mo),x2(x(Mo))]
4142                                  >Sith[Io,x(Io)]; Fin[Mo,x(Mo)] end
4143                              >Sith[Io1]; Fin[Mo1] end
4144                              >Sith[Io2]; Fin[Mo2]
4145                                  >Sith[Io,AdCmax(Io)]; Fin[Mo,AdCmax(Mo)]
4146                                      >Sith[Io,AdCmax(Io),x2(AdCmax(Io))];
4147                                      >Fin[Mo,AdCmax(Mo),x2(AdCmax(Mo))]
4148                                  >Sith[Io,AdCmax(Io)]; Fin[Mo,AdCmax(Mo)] end
4149                                  >Sith[C_AdC]
4150                                      >Sith[C_AdC,w_AdC]
4151                                  >Sith[C_AdC]
4152                                  >Sith[G_AdC]
4153                                      >Sith[Subject_AdC]
4154                                          >Sith[Subject_AdC,
4155                                          >d_AdC(Subject_AdC)]
4156                                              >Sith[Subject_AdC,
4157                                              >d_AdC(Subject_AdC),
4158                                              >c_AdC(d_AdC(Subject_AdC))]
4159                                          >Sith[Subject_AdC,
4160                                          >d_AdC(Subject_AdC)] end
4161                                      >Sith[Subject_AdC] end
4162                                      >Sith[Object_AdC]
4163                                          >Sith[Object_AdC,
4164                                          >d_AdC(Object_AdC)]
4165                                              >Sith[Object_AdC,
4166                                              >d_AdC(Object_AdC),
4167                                              >c_AdC(d_AdC(Object_AdC))]
4168                                          >Sith[Object_AdC,
4169                                          >d_AdC(Object_AdC)] end
4170                                      >Sith[Object_AdC] end
4171                                  >Sith[G_AdC] end
4172                              >Sith[Io2]; Fin[Mo2] end
4173                          >Sith[To]; Fin[To] end
4174                      >Sith[T]; Fin[T] end
4175                  >structure end
4176                  >Konw[Conc(C_AdC,G_AdC)]
4177                   >E[Ч,e[Ч](C_AdC)]
4178                   >E[Ч,e[Ч](G_AdC)]
4179                  >Konw[Conc(C_AdC,G_AdC)]end
4180                  >Konw[Causality_AdC]
4181                   >of Sith[Subject_AdC], Sith[Subject_AdC,d_AdC(Subject_AdC)],
4182                   >Sith[Subject_AdC,d_AdC(Subject_AdC),c_AdC(d_AdC(Subject_AdC))]
4183                   >Sith[Object_AdC], Sith[Object_AdC,d_AdC(Object_AdC)],
4184                   >Sith[Object_AdC,d_AdC(Object_AdC),c_AdC(d_AdC(Object_AdC))]
4185                  >Konw[Causality_AdC]end
4186                  >Konp[Obs5A_AdC_4(i_AdC)]
4187                   >of Sith[Obs5A_AdC_4(iAdC)], Fin[Obs]
4188                   >E[Ц,e[Ц](5A_AdC_4(i_AdC))], E[Ц,e[Ц](R5)]
4189                  >Konp[Obs5A_AdC_4(i_AdC)]end
4190                  >Konr[Descript((T,(G_AdC)),R5)]
4191                   >E[Ч,e[Ч](T)], E[Ч,e[Ч](G_AdC))]
4192                   >E[Ц,e[Ц](R5)]
4193                  >Konr[Descript((T,(G_AdC)),R5)]
```

```
4194                      >Konr[Prior,Sits[R5]]
4195                       >Fin[Mo,x(Mo,p)]
4196                       >Fin[Mo,x(Mo,p)+1]
4197                      >Konr[Prior,Sits[R5]]end
4198                    >Sits[5A_AdC_4(i_AdC)]end
4199            >Kont[AdC(i_AdC)]end
4200            >Kont[ReCore]
4201             >Sits[5A_ReC_1]=Sits[5A(1[0])]
4202              >E[Ц,e[Ц](5A_ReC_1)]=E[Ц,e[Ц](5A(1[0]))]
4203                  >Sits[5A_ReC_1]
4204                    >structure
4205                        >Sits[R5]
4206                         >E[Ц,e[Ц](R5)]
4207                        >Sith[Obs5A_ReCore_1]; Fin[Obs]
4208                        >Sith[T]; Fin[T]
4209                           >Sith[To]; Fin[To]
4210                              >Sith[Io,x(Io)]; Fin[Mo,x(Mo)]
4211                                 >Sith[Io,x(Io),x2(x(Io))];
4212                                 >Fin[Mo,x(Mo),x2(x(Mo))]
4213                              >Sith[Io,x(Io)]; Fin[Mo,x(Mo)] end
4214                           >Sith[To]; Fin[To] end
4215                        >Sith[T]; Fin[T] end
4216                    >structure end
4217                    >Konp[Obs5A_ReC_1]
4218                     >of Sith[Obs5A_ReC_1], Fin[Obs]
4219                     >E[Ц,e[Ц](5A_ReC_1)], E[Ц,e[Ц](R5)]
4220                    >Konp[Obs5A_ReC_1]end
4221                    >Konr[Descript(T,R5)]
4222                     >E[Ч,e[Ч](T)]
4223                     >E[Ц,e[Ц](R5)]
4224                    >Konr[Descript(T,R5)]end
4225                    >Konr[Prior,Sits[R5]]
4226                     >Fin[Mo,x(Mo,p)]
4227                     >Fin[Mo,x(Mo,p)+1]
4228                    >Konr[Prior,Sits[R5]]end
4229                  >Sits[5A_ReC_1]end
4230            >Konw[4ReCore_2]
4231             >of Sits[5A_NiB_1]
4232             >Sits[5A_NiB_2]
4233             >x(Mo)[1:p(Mo)-1]
4234             >x(Mo,p)[1:p(Mo)-2]
4235             >Fin[Mo,x(Mo)]
4236              >\Fin[Mo,x(Mo,p)max+1]
4237             >p(Mo)=p(Mo)-1
4238             >x(Mo)[1:p(Mo)]
4239             >x(Mo,p)[1:p(Mo)-1]
4240            >Konw[4ReCore_2]end
4241                  >Sits[5A_ReC_2]
4242                    >structure
4243                        >Sits[R5]
4244                         >E[Ц,e[Ц](R5)]
4245                        >Sith[Obs5A_ReC_2]; Fin[Obs]
4246                        >Sith[T]; Fin[T]
4247                           >Sith[To]; Fin[To]
4248                              >Sith[Io,x(Io)]; Fin[Mo,x(Mo)]
4249                                 >Sith[Io,x(Io),x2(x(Io))];
4250                                 >Fin[Mo,x(Mo),x2(x(Mo))]
4251                              >Sith[Io,x(Io)]; Fin[Mo,x(Mo)] end
4252                           >Sith[To]; Fin[To] end
4253                        >Sith[T]; Fin[T] end
```

71

```
4254            >structure end
4255            >Konp[Obs5A_ReC_2]
4256             >of Sith[Obs5A_ReC_2], Fin[Obs]
4257             >E[Ц,e[Ц](5A_ReC_2)], E[Ц,e[Ц](R5)]
4258            >Konp[Obs5A_ReC_2]end
4259            >Konr[Descript(T,R5)]
4260             >E[Ч,e[Ч](T)]
4261             >E[Ц,e[Ц](R5)]
4262            >Konr[Descript(T,R5)]end
4263            >Konr[Prior,Sits[R5]]
4264             >Fin[Mo,x(Mo,p)]
4265             >Fin[Mo,x(Mo,p)+1]
4266            >Konr[Prior,Sits[R5]]end
4267          >Sits[5A_ReC]end
4268      >Konw[4ReCore_3]
4269       >of Sits[5A_ReC_2]
4270       >Sith[C1_ReC,c_ReC]=mutual(Sith[R5,a_ReC,b_ReC(a_ReC)])
4271       >while (c_ReC=0):
4272         >\Sits[R5]
4273        >Sits[R5new]\=Sits[R5]
4274            >Sits[R5new]
4275             >structure
4276                >Sith[R5new]; Fin[R5new]
4277                     >Sith[R5new,a_ReCnew]; Fin[R5new,a_ReCnew]
4278                          >Sith[R5new,a_ReCnew,b_ReCnew(a_ReCnew)]
4279                     >Sith[R5new,a_ReCnew]; Fin[R5new,a_ReCnew] end
4280                >Sith[R5new]; Fin[R5new] end
4281             >strucutre end
4282            >Sit[R5new] end
4283        >Sits[R5]=Sits[R5new]
4284         >E[Ц,e[Ц](R5)=E[Ц,e[Ц](R5new)
4285       >else:
4286            >Sits[C1_ReC]
4287             >structure
4288                >Sith[C1_ReC]
4289                     >Sith[C1_ReC,c_ReC]
4290                >Sith[C1_ReC] end
4291             >structure end
4292            >Sits[C1_ReC] end
4293        >Sith[d_ReC]=Classes(Sith[C1_ReC])
4294            >Sits[C2_ReC]
4295             >structure
4296                >Sith[d_ReC]
4297                       >Sith[d_ReC,c_ReC(d_ReC)]
4298                >Sith[d_ReC] end
4299             >structure end
4300            >Sits[C2_ReC] end
4301        >Sith[C_ReC]=Konw[M](Sits[C2_ReC])
4302         >E[Ч,e[Ч](C_ReC)]=E[3,e[3](C2_ReC)]
4303        >Sits[5A_ReC_3]
4304       >endloop
4305      >Konw[4ReCore_3]end
4306            >Sits[5A_ReC_3]
4307             >structure
4308                >Sits[R5]
4309                 >E[Ц,e[Ц](R5)]
4310                >Sits[Ex4oC_ReC]
4311                 >E[Ц,e[Ц](Ex4oC_ReC)]
4312                >Sith[Obs5A_ReC_3]; Fin[Obs]
4313                >Sith[T]; Fin[T]
```

```
4314                              >Sith[To]; Fin[To]
4315                                  >Sith[Io1]; Fin[Mo1]
4316                                      >Sith[Io,x(Io)]; Fin[Mo,x(Mo)]
4317                                          >Sith[Io,x(Io),x2(x1(Io))];
4318                                          >Fin[Mo,x(Mo),x2(x1(Mo))]
4319                                      >Sith[Io,x(Io)]; Fin[Mo,x(Mo)] end
4320                                  >Sith[Io1]; Fin[Mo1] end
4321                                  >Sith[Io2]; Fin[Mo2]
4322                                      >Sith[C_ReC]
4323                                          >Sith[C_ReC,w_ReC]
4324                                      >Sith[C_ReC] end
4325                                  >Sith[Io2]; Fin[Mo2] end
4326                              >Sith[To]; Fin[To] end
4327                      >Sith[T]; Fin[T] end
4328              >structure end
4329              >Konp[Obs5A_ReC_3]
4330               >of Sith[Obs5A_Re_3], Fin[Obs]
4331               >E[Ч,e[Ц](5A_ReC_3)], E[Ц,e[Ц](R5)]
4332              >Konp[Obs5A_ReC_3]end
4333              >Konr[Descript(T,R5)]
4334               >E[Ч,e[Ч](T)]
4335               >E[Ц,e[Ц](R5)]
4336              >Konr[Descript(T,R5)]end
4337              >Konr[Prior,Sits[R5]]
4338               >Fin[Mo,x(Mo,p)]
4339               >Fin[Mo,x(Mo,p)+1]
4340              >Konr[Prior,Sits[R5]]end
4341             >Sits[5A_ReC_3]end
4342      >Konw[4ReCore_4]
4343       >of Sits[5A_ReC_3]
4344       >w_ReC1,w_ReC2=groups(w_rEC)
4345      >w_ReC1\=w_ReC2
4346      >Konr[w_ReC1,w_ReC2]
4347        >Sith[C,w_ReC1]
4348        >Sith[C,w_ReC2]
4349       >Konr[w_ReC1,w_ReC2]end
4350      >Sits[G_ReC]=Konf[int](Konr[w_ReC1,w_ReC2])
4351      >E[Ч,e[Ч](Subject_ReC)]=Sith[C,w_ReC1]
4352       >E[Ч,e[Ч](Object_ReC)]=Sith[C,w_ReC2]
4353      >Konw[Causality_ReC]=Class(Konr[w_ReC1,w_ReC2])
4354           >Sits[G_ReC]
4355            >structure
4356               >Sith[G_ReC]
4357                   >Sith[Subject_ReC]
4358                       >Sith[Subject_ReC,d_ReC(Subject_ReC)]
4359                           >Sith[Subject_ReC,d_ReC(Subject_ReC),
4360                           >c_ReC(d_ReC(Subject_ReC))]
4361                       >Sith[Subject_ReC,d_ReC(Subject_ReC)] end
4362                   >Sith[Subject_ReC] end
4363                   >Sith[Object_ReC]
4364                       >Sith[Object_ReC,d_ReC(Object_ReC)]
4365                           >Sith[Object_ReC,d_ReC(Object_ReC),
4366                           >c_ReC(d_ReC(Object_ReC))]
4367                       >Sith[Object_ReC,d_ReC(Object_ReC)] end
4368                   >Sith[Object_ReC] end
4369               >Sith[G_ReC] end
4370            >structure end
4371            >Konw[Causality_ReC]
4372             >of Sith[Subject_ReC], Sith[Subject_ReC,d_ReC(Subject_ReC)],
4373             >Sith[Subject_ReC,d_ReC(Subject_ReC),c_ReC(d_ReC(Subject_ReC))]
```

```
4374              >Sith[Object_ReC], Sith[Object_ReC,d_ReC(Object_ReC)],
4375                 >Sith[Object_ReC,d_ReC(Object_ReC),c_ReC(d_ReC(Object_ReC))]
4376           >Konw[Causality_ReC]end
4377          >Sits[G_ReC] end
4378     >Sith[C_ReC]=Konw[M](Sits[G_ReC])
4379      >E[Ч,e[Ч](G_ReC)]=E[3,e[3](G_ReC)]
4380     >Sits[5A_ReC_4]
4381  >Konw[4ReCore_4] end
4382            >Sits[5A_ReC_4]
4383             >structure
4384                 >Sits[R5]
4385                  >E[Ц,e[Ц](R5)]
4386                 >Sith[Obs5A_ReC-4]; Fin[Obs]
4387                 >Sith[T]; Fin[T]
4388                     >Sith[To]; Fin[To]
4389                         >Sith[Io1]; Fin[Mo1]
4390                             >Sith[Io,x(Io)]; Fin[Mo,x(Mo)]
4391                                 >Sith[Io,x(Io),x2(x1(Io))];
4392                                   >Fin[Mo,x(Mo),x2(x1(Mo))]
4393                             >Sith[Io,x(Io)]; Fin[Mo,x(Mo)] end
4394                         >Sith[Io1]; Fin[Mo1] end
4395                         >Sith[Io2]; Fin[Mo2]
4396                             >Sith[C_ReC]
4397                                 >Sith[C_ReC,w_ReC]
4398                             >Sith[C_ReC] end
4399                             >Sith[G_ReC]
4400                                 >Sith[Subject_ReC]
4401                                     >Sith[Subject_ReC,
4402                                     >d_ReC(Subject_ReC)]
4403                                         >Sith[Subject_ReC,
4404                                         >d_ReC(Subject_ReC),
4405                                         >c_ReC(d_ReC(Subject_ReC))]
4406                                     >Sith[Subject_ReC,
4407                                     >d_ReC(Subject_ReC)] end
4408                                 >Sith[Subject_ReC] end
4409                                 >Sith[Object_ReC]
4410                                     >Sith[Object_ReC,
4411                                     >d_ReC(Object_ReC)]
4412                                         >Sith[Object_ReC,
4413                                         >d_ReC(Object_ReC),
4414                                         >c_ReC(d_ReC(Object_ReC))]
4415                                     >Sith[Object_ReC,
4416                                     >d_ReC(Object_ReC)] end
4417                                 >Sith[Object_ReC] end
4418                             >Sith[G_ReC] end
4419                         >Sith[Io2]; Fin[Mo2] end
4420                     >Sith[To]; Fin[To] end
4421                 >Sith[T]; Fin[T] end
4422             >structure end
4423             >Konw[Conc(C_ReC,G_ReC)]
4424              >E[Ч,e[Ч](C_ReC)]
4425              >E[Ч,e[Ч](G_ReC)]
4426             >Konw[Conc(C_ReC,G_ReC)]end
4427             >Konw[Causality_ReC]
4428              >of Sith[Subject_ReC], Sith[Subject_ReC,d_ReC(Subject_ReC)],
4429              >Sith[Subject_ReC,d_ReC(Subject_ReC),c_ReC(d_ReC(Subject_ReC))]
4430              >Sith[Object_ReC], Sith[Object_ReC,d_ReC(Object_ReC)],
4431              >Sith[Object_ReC,d_ReC(Object_ReC),c_ReC(d_ReC(Object_ReC))]
4432             >Konw[Causality_ReC]end
4433             >Konp[Obs5A_ReC_4]
```

```
4434              >of Sith[Obs5A_ReC_4], Fin[Obs]
4435               >E[Ц,e[Ц](5A_ReC_4)], E[Ц,e[Ц](R5)]
4436              >Konp[Obs5A_ReC_4]end
4437             >Konr[Descript(T,R5)]
4438              >E[Ч,e[Ч](T)]
4439              >E[Ц,e[Ц](R5)]
4440             >Konr[Descript(T,R5)]end
4441             >Konr[Prior,Sits[R5]]
4442              >Fin[Mo,x(Mo,p)]
4443              >Fin[Mo,x(Mo,p)+1]
4444             >Konr[Prior,Sits[R5]]end
4445            >Sits[5A_ReC_4]end
4446       >Konf[Ex_ReC]
4447        >of Sits[5A_ReC_4]
4448        >Konr[Descript(G_ReC,R5)]
4449         >E[Ч,e[Ч](G_ReC)]
4450         >E[Ц,e[Ц](R5)]
4451        >Konr[Descript(G_ReC,R5)]end
4452        >if (Konr[Descript(G_ReC,R5)]=omnivial):
4453         >ExReCA=1
4454        >else:
4455         >ExReCA=0
4456        >endif
4457        >i_ReC=1
4458        >ExReCloop=1
4459        >while (ExReCloop=1):
4460         >if (ExReCA=1):
4461            >Sits[Modelo_ReC]
4462             >structure
4463                 >Sith[chi]; Fin[chi]
4464                    >Sith[Subject_ReC]
4465                        >Sith[Subject_ReC,d_ReC(Subject_ReC)]
4466                            >Sith[Subject_ReC,d_ReC(Subject_ReC),
4467                             >c_ReC(d_ReC(Subject_ReC))]
4468                        >Sith[Subject_ReC,d_ReC(Subject_ReC)] end
4469                    >Sith[Subject_ReC] end
4470                    >Sith[Object_ReC]
4471                        >Sith[Object_ReC,d_ReC(Object_ReC)]
4472                            >Sith[Object_ReC,d_ReC(Object_ReC),
4473                             >c_ReC(d_ReC(Object_ReC))]
4474                        >Sith[Object_ReC,d_ReC(Object_ReC)] end
4475                    >Sith[Object_ReC] end
4476                 >Sith[chi]; Fin[chi] end
4477             >structure end
4478             >Konw[NG_ReC]
4479              >of Sith[Subject_ReC], Sith[Subject_ReC,d_ReC(Subject_ReC)],
4480              >Sith[Subject_ReC,d_ReC(Subject_ReC),c_ReC(d_ReC(Subject_ReC))]
4481              >Sith[Object_ReC], Sith[Object_ReC,d_ReC(Object_ReC)],
4482              >Sith[Object_ReC,d_ReC(Object_ReC),c_ReC(d_ReC(Object_ReC))]
4483             >Konw[NG_ReC] end
4484            >Sits[Modelo_ReC]] end
4485            >Sits[h,Io_ReC]]
4486             >structure
4487                 >Sith[chi]; Fin[chi]
4488                    >Sith[Subject_ReC]
4489                        >Sith[Subject_ReC,d_ReC(Subject_ReC)]
4490                            >Sith[Subject_ReC,d_ReC(Subject_ReC),
4491                             >c_ReC(d_ReC(Subject_ReC))]
4492                        >Sith[Subject_ReC,d_ReC(Subject_ReC)] end
4493                    >Sith[Subject_ReC] end
```

```
4494                            >Sith[chi]; Fin[chi] end
4495                      >structure end
4496                    >Sits[h,Io_ReC] end
4497              >Class(Sith[chi],Fin[chi](Sits[h,Io,_ReC]))=
4498          >Konf[M](E[Ц,e[Ц](Modelo_ReC)],Konw[NG_ReC])
4499          >Konr[Subject_ReC,chi]
4500            >Sith[Subject_ReC]
4501            >Sith[chi]
4502           >Konr[Subject_ReC,chi]end
4503          >Konr[Subject_ReC,chi]=Konf[M](Konw[NG_ReC])
4504          >Class(Konr[Subject_ReC,chi])=Ef
4505          >Sith[Io_ReC] Sith[Io_ReC,x_ReC(Io_ReC)]
4506           >E[Ч^5,e[Ч^5](Io_ReC)]=Class(Sith[chi],Fin[chi])
4507          >Fin[Mo_ReC], Fin[Mo_ReC,x_ReC(Mo_ReC)]=Konw[M](Sits[Modelo_ReC])
4508           >E[Ч^5,e[Ч^5](Mo_ReC)]=E[3,e[3](Modelo_ReC)]
4509          >q_ReC(Io_ReC)=abs(Class(Sith[chi],Fin[chi]))
4510          >q_ReC(Mo_ReC)=abs(ALL(E[3,e[3](Modelo_ReC)]))
4511          >Sits[5A_ReC_5]
4512        >elif (ExReCA=0):
4513        >s_ReC[1:i_ReC]
4514        >w_ReC2(i_ReC)=w_ReC1
4515        >w_ReC2(i_ReC)=w_ReC2
4516        >Sith[i_ReC,Subject_ReC(i_ReC),d(Subject_ReC(i_ReC)),
4517        >c_ReC(d_ReC(Subject_ReC(i_ReC)))]=
4518        >Sith[Subject_ReC,d_ReC(Subject_ReC),c_ReC(d_ReC(Subject_ReC))]
4519        >Sith[i_ReC,Subject_ReC,d_ReC(Subject(i_ReC))]=
4520        >Sith[Subject_ReC,d_ReC(Subject_ReC)]
4521        >Sith[i_ReC,Subject_ReC(i_ReC)]=Sith[Subject_ReC]
4522        >Sith[i_ReC,Object_ReC(i_ReC),d(Object_ReC(i_ReC)),
4523        >c_ReC(d_ReC(Object_ReC(i_ReC)))]=
4524        >Sith[Object_ReC,d_ReC(Object_ReC),c_ReC(d_ReC(Object_ReC))]
4525        >Sith[i_ReC,Object_ReC,d_ReC(Object(i_ReC))]=Sith[Object_ReC,d_ReC(Object_ReC)]
4526        >Sith[i_ReC,Object_ReC(i_ReC)]=Sith[Object_ReC]
4527            >Sits[F(i_ReC)]
4528             >structure
4529                >Sith[Guess,s_ReC]
4530                    >Sith[s_ReB,Subject(s_ReC)]
4531                        >Sith[s_ReC,Subject_ReC(s_ReC),d_ReC(Subject(s_ReC))]
4532                            >Sith[s_ReC,Subject_ReC(s_ReC),
4533                            >d_ReC(Subject_ReC(s_ReC)),
4534                            >c_ReC(d_ReC(Subject_ReC(s_ReC)))]
4535                        >Sith[s_ReC,Subject_ReC(s_ReC),
4536                        >d_ReC(Subject_ReC(s_ReC))] end
4537                    >Sith[s_ReC,Subject(s_ReC)] end
4538                    >Sith[s_ReC,Object(s_ReC)]
4539                        >Sith[s_ReC,Object_ReC(s_ReC),d_ReC(Object_ReC(s_ReC))]
4540                            >Sith[s_ReC,Object_ReC(s_ReC),
4541                            >d_ReB(Object_ReC(s_ReC)),
4542                            >c_ReC(d_ReC(Object_ReC(s_ReC)))]
4543                        >Sith[s_ReC,Object_ReC(s_ReC),
4544                        >d_ReC(Object_ReC(s_ReC))] end
4545                    >Sith[s_ReC,Object_ReC(s_ReC)] end
4546                >Sith[Guess,s_ReC] end
4547            >structure end
4548            >Konw[Causality_ReC]
4549             >of Sith[s_ReC,Subject_ReC(s_ReC)],
4550             >Sith[s_ReC,Subject_ReC(s_ReC),d_ReC(Subject_ReC(s_ReC))],
4551             >Sith[Subject_ReC(s_ReC),d_ReC(Subject_ReC(s_ReC)),
4552             >c_ReC(d_ReC(Subject_ReC(s_ReC)))]
```

```
4553          >Sith[s_ReC,Object_ReC(s_ReC)],
4554          >Sith[s_ReC,Object_ReC(s_ReC),d_ReC(Object_ReC(s_ReC))],
4555          >Sith[s_ReC,Object_ReC(s_ReC),d_ReC(Object_ReC(s_ReC)),
4556          >c_ReC(d_ReC(Object_ReC(s_ReC)))]
4557        >Konw[Causality_ReC]end
4558       >Sits[F(i_ReC)]end
4559     >w_ReC1,w_ReC2=groups(m_ReC)
4560      >w_ReC1\=w_ReC1(s_ReC)
4561      >w_ReC2\=w_ReC2(s_ReC)
4562     >Konr[w_ReC1,w_ReC2]
4563       >Sith[C,w_ReC1]
4564       >Sith[C,w_ReC2]
4565      >Konr[w_ReC1,w_ReC2]end
4566      >Sits[G]=Konf[int](Konr[w_ReC1,w_ReC2])
4567      >E[Ч,e[Ч](Subject_ReC)]=Sith[C,w_ReC1]
4568       >E[Ч,e[Ч](Object_ReC)]=Sith[C,w_ReC2]
4569     >Konw[Causality_ReC]=Class(Konr[w_ReC1,w_ReC2])
4570     >Sits[G_ReC]
4571     >E[Ц,e[Ц](G_ReC)]\=E[Ц,e[Ц](F(i_ReC))]
4572     >if (abs(Sits[G_ReC])\=0):
4573      >ExReCloop=1
4574     >elif (abs(Sits[G_ReC])=0):
4575      >ExReCloop=0
4576      >proceed
4577     >endif
4578         >Sits[G_ReC]
4579          >structure
4580            >Sith[Subject_ReC]
4581                 >Sith[Subject_ReC,d_ReC(Subject_ReC)]
4582                    >Sith[Subject_ReC,d_ReC(Subject_ReC),
4583                    >c_ReC(d_ReC(Subject_ReC))]
4584                 >Sith[Subject_ReC,d_ReC(Subject_ReC)] end
4585            >Sith[Subject_ReC] end
4586            >Sith[Object_ReC]
4587                 >Sith[Object_ReC,d_ReC(Object_ReC)]
4588                    >Sith[Object_ReC,d_ReC(Object_ReC),
4589                    >c_ReC(d_ReC(Object_ReC))]
4590                 >Sith[Object_ReC,d_ReC(Object_ReC)] end
4591            >Sith[Object_ReC] end
4592         >structure end
4593         >Konw[Causality_ReC]
4594          >of Sith[Subject_ReC], Sith[Subject_ReC,d_ReC(Subject_ReC)],
4595          >Sith[Subject_ReC,d_ReC(Subject_ReC),c_ReC(d_ReC(Subject_ReC))]
4596          >Sith[Object_ReC], Sith[Object_ReC,d_ReC(Object_ReC)],
4597          >Sith[Object_ReC,d_ReC(Object_ReC),c_ReC(d_ReC(Object_ReC))]
4598         >Konw[Causality_ReC]end
4599        >Sits[G_ReC] end
4600      >Sith[C_ReC]=Konw[M](Sits[G_ReC])
4601       >E[Ч,e[Ч](G_ReC)]=E[3,e[3](G_ReC)]
4602      >Kont[ReC(i_ReC)]
4603      >i_ReC=i_ReC+1
4604     >endif
4605     >else
4606     >Sits[5A_ReC_1]
4607     >Sits[5A_ReC]=Sits[5A_ReC_1]
4608      >E[Ц,e[Ц](5A_ReC)]=E[Ц,e[Ц](5A_ReC_1)]
4609     >endloop
4610   >Konf[Ex_ReC]end
4611         >Sits[5A_ReC_5]
4612          >structure
```

```
4613                          >Sits[R5]
4614                           >E[Ц,e[Ц](R5)]
4615                          >Sith[Obs5A_ReC_5]; Fin[Obs]
4616                          >Sith[T]; Fin[T]
4617                              >Sith[To]; Fin[To]
4618                                  >Sith[Io1]; Fin[Mo]
4619                                      >Sith[Io,x(Io)]; Fin[Mo,x(Mo)]
4620                                          >Sith[Io,x(Io),x2(x1(Io))];
4621                                          >Fin[Mo,x(Mo),x2(x1(Mo))]
4622                                      >Sith[Io,x(Io)]; Fin[Mo,x(Mo)] end
4623                                  >Sith[Io1]; Fin[Mo1] end
4624                                      >Sith[Io2]
4625                                          >Sith[Io_AdC]; Fin[Mo_AdC]
4626                                              >Sith[Io_AdC,x_AdC(Io_AdC)];
4627                                              >Fin[Mo_AdC,x_AdC(Mo_AdC)]
4628                                          >Sith[Io_AdC]; Fin[Mo_AdC] end
4629                                      >Sith[Io2] end
4630                                  >Sith[Io1]; Fin[Mo1] end
4631                              >Sith[To]; Fin[To] end
4632                          >Sith[T]; Fin[T] end
4633                  >structure end
4634                  >Konp[Obs5A_ReC_5]
4635                   >of Sith[Obs5A_ReC_5], Fin[Obs]
4636                   >E[Ц,e[Ц](5A_ReC_5)], E[Ц,e[Ц](R5)]
4637                  >Konp[Obs5A_ReC_5]end
4638                  >Konr[Descript(T,R5)]
4639                   >E[Ч,e[Ч](T)]
4640                   >E[Ц,e[Ц](R5)]
4641                  >Konr[Descript(T,R5)]end
4642                  >Konr[Prior,Sits[R5]]
4643                   >Fin[Mo,x(Mo,p)]
4644                   >Fin[Mo,x(Mo,p)+1]
4645                  >Konr[Prior,Sits[R5]]end
4646              >Sits[5A_ReC_5]end
4647      >Konw[5A_ReB_6]
4648       >of Sits[5A_ReB_5]
4649      >Fin[Mo,x(Mo),x2(x(Mo))]=Fin[Mo,x(Mo),x2(x(Mo))]+Fin[Mo_ReC,x_ReC(Mo_ReC)]
4650      >Fin[Mo,x(Mo)]=Fin[Mo,x(Mo)]+Fin[Mo_ReC]
4651      >p(Mo)=p(Mo)-1
4652      >x(Mo)[1:p(Mo)]
4653      >x(Mo,p)[1:p(Mo)-1]
4654      >Sits[5A_ReC_6]
4655      >Konw[5A_ReB_6] end
4656              >Sits[5A_ReC_6]
4657               >structure
4658                  >Sits[R5]
4659                   >E[Ц,e[Ц](R5)]
4660                  >Sith[Obs5A_ReC_6]; Fin[Obs]
4661                  >Sith[T]; Fin[T]
4662                      >Sith[To]; Fin[To]
4663                          >Sith[Io,x(Io)]; Fin[Mo,x(Mo)]
4664                              >Sith[Io,x(Io),x2(x1(Io))];
4665                              >Fin[Mo,x(Mo),x2(x1(Mo))]
4666                          >Sith[Io,x(Io)]; Fin[Mo,x(Mo)] end
4667                      >Sith[To]; Fin[To] end
4668                  >Sith[T]; Fin[T] end
4669              >structure end
4670              >Konp[Obs5A_ReC_6]
4671               >of Sith[Obs5A_ReC_6], Fin[Obs]
4672               >E[Ц,e[Ц](5A_ReC_6)], E[Ц,e[Ц](R5)]
```

```
4673              >Konp[Obs5A_ReC_6]end
4674              >Konr[Descript(T,R5)]
4675               >E[Ч,e[Ч](T)]
4676               >E[Ц,e[Ц](R5)]
4677              >Konr[Descript(T,R5)]end
4678              >Konr[Prior,Sits[R5]]
4679               >Fin[Mo,x(Mo,p)]
4680               >Fin[Mo,x(Mo,p)+1]
4681              >Konr[Prior,Sits[R5]]end
4682           >Sits[5A_ReC_6]end
4683       >Sits[5A_ReC]=Sits[5A_ReC_6]
4684        >E[Ц,e[Ц](5A_ReC)]=E[Ц,e[Ц](5A_ReC_6)]
4685     >Kont[ReCore]end
4686     >Kont[ReC(i_ReC)]
4687            >Sits[5A_ReC_4(i_ReC)]
4688             >structure
4689                  >Sits[R5]
4690                   >E[Ц,e[Ц](R5)]
4691                  >Sith[Obs5A_ReC-4(i_ReC)]; Fin[Obs]
4692                  >Sith[T]; Fin[T]
4693                     >Sith[To]; Fin[To]
4694                        >Sith[Io1]; Fin[Mo1]
4695                           >Sith[Io,x(Io)]; Fin[Mo,x(Mo)]
4696                              >Sith[Io,x(Io),x2(x1(Io))];
4697                              >Fin[Mo,x(Mo),x2(x1(Mo))]
4698                           >Sith[Io,x(Io)]; Fin[Mo,x(Mo)] end
4699                        >Sith[Io1]; Fin[Mo1] end
4700                        >Sith[Io2]; Fin[Mo2]
4701                           >Sith[C_ReC]
4702                              >Sith[C_ReC,w_ReC]
4703                           >Sith[C_ReC] end
4704                           >Sith[G_ReC]
4705                              >Sith[Subject_ReC]
4706                                 >Sith[Subject_ReC,
4707                                 >d_ReC(Subject_ReC)]
4708                                    >Sith[Subject_ReC,
4709                                    >d_ReC(Subject_ReC),
4710                                    >c_ReC(d_ReC(Subject_ReC))]
4711                                 >Sith[Subject_ReC,
4712                                 >d_ReC(Subject_ReC)] end
4713                              >Sith[Subject_ReC] end
4714                              >Sith[Object_ReC]
4715                                 >Sith[Object_ReC,
4716                                 >d_ReC(Object_ReC)]
4717                                    >Sith[Object_ReC,
4718                                    >d_ReC(Object_ReC),
4719                                    >c_ReC(d_ReC(Object_ReC))]
4720                                 >Sith[Object_ReC,
4721                                 >d_ReC(Object_ReC)] end
4722                              >Sith[Object_ReC] end
4723                           >Sith[G_ReC] end
4724                        >Sith[Io2]; Fin[Mo2] end
4725                     >Sith[To]; Fin[To] end
4726                  >Sith[T]; Fin[T] end
4727             >structure end
4728             >Konw[Conc(C_ReC,G_ReC)]
4729              >E[Ч,e[Ч](C_ReC)]
4730              >E[Ч,e[Ч](G_ReC)]
4731             >Konw[Conc(C_ReC,G_ReC)]end
4732             >Konw[Causality_ReC]
```

```
4733            >of Sith[Subject_ReC], Sith[Subject_ReC,d_ReC(Subject_ReC)],
4734            >Sith[Subject_ReC,d_ReC(Subject_ReC),c_ReC(d_ReC(Subject_ReC))]
4735            >Sith[Object_ReC], Sith[Object_ReC,d_ReC(Object_ReC)],
4736            >Sith[Object_ReC,d_ReC(Object_ReC),c_ReC(d_ReC(Object_ReC))]
4737        >Konw[Causality_ReC]end
4738        >Konp[Obs5A_ReC_4(i_ReC)]
4739            >of Sith[Obs5A_ReC_4(i_ReC)], Fin[Obs]
4740            >E[Ц,e[Ц](5A_ReC_4(i_ReC))], E[Ц,e[Ц](R5)]
4741        >Konp[Obs5A_ReC_4(i_ReC)]end
4742        >Konr[Descript((T,(G_ReC)),R5)]
4743            >E[Ч,e[Ч](T)], E[Ч,e[Ч](G_ReC)
4744            >E[Ц,e[Ц](R5)]
4745        >Konr[Descript((T,(G_ReC)),R5)]end
4746        >Konr[Prior,Sits[R5]]
4747            >Fin[Mo,x(Mo,p)]
4748            >Fin[Mo,x(Mo,p)+1]
4749        >Konr[Prior,Sits[R5]]end
4750        >Sits[5A_ReC_4(i_ReC)]end
4751    >Kont[ReC(i_ReC)]end
4752    >Kont[ordBelt]
4753        >Sits[ordBelt]
4754        >structure
4755            >Sits[5A_NiB]
4756            >E[Ц,e[Ц](5A_NiB)]
4757                >Sith[T], Fin[T]
4758                >E[Ч,e[Ч](5A_NiB_T)
4759            >Konr[Descript(T,R5)]
4760                >Sith[T], Fin[T]
4761                >E[Ц,e[Ц](R5)]
4762            >Konr[Descript(T,R5)]end
4763            >Konf[ordBelt_NiB]
4764                >of Konr[Descript(T,R5)]
4765                >NiB=abs(viality(Konr[Descript(T,R5)]))
4766            >Konf[ordBelt_NiB]end
4767            >Sits[5A_NiB]end
4768            >Sits[5A_AdB]
4769            >E[Ц,e[Ц](5A_AdB)]
4770                >Sith[T], Fin[T]
4771                >E[Ч,e[Ч](5A_AdB_T)
4772            >Konr[Descript(T,R5)]
4773                >Sith[T], Fin[T]
4774                >E[Ц,e[Ц](R5)]
4775            >Konr[Descript(T,R5)]end
4776            >Konf[ordBelt_AdB]
4777                >of Konr[Descript(T,R5)]
4778                >AdB=abs(viality(Konr[Descript(T,R5)]))
4779            >Konf[ordBelt_AdB]end
4780            >Sits[5A_AdB]end
4781            >Sits[5A_ReB]
4782            >E[Ц,e[Ц](5A_ReB)]
4783                >Sith[T], Fin[T]
4784                >E[Ч,e[Ч](5A_ReB_T)
4785            >Konr[Descript(T,R5)]
4786                >Sith[T], Fin[T]
4787                >E[Ц,e[Ц](R5)]
4788            >Konr[Descript(T,R5)]end
4789            >Konf[ordBelt_ReB]
4790                >of Konr[Descript(T,R5)]
4791                >ReB=abs(viality(Konr[Descript(T,R5)]))
4792            >Konf[ordBelt_ReB]end
```

```
4793                         >Sits[5A_ReB]end
4794                 >structure end
4795                 >Konr[ordBelt1]
4796                  >XBmin
4797                  >XBmid
4798                  >XBmax
4799                 >Konr[ordBelt1]end
4800                 >Konf[ordBelt2]
4801                  >of XBmin
4802                  >if (XBmax=NiB) then
4803                   >Sits[5A_NiB]
4804                    >E[Ц,e[Ц](5A_NiB)]
4805                    >\Konr[Descript(T,R5)]
4806                    >\Konf[ordBelt_NiB]
4807                   >Sits[5A_BPrior]=Sits[5A_NiB]
4808                    >E[Ц,e[Ц](5A_BPrior)]=E[Ц,e[Ц](5A_NiB)]
4809                  >elif (XBmax=AdB) then
4810                   >Sits[5A_AdB]
4811                    >E[Ц,e[Ц](5A_AdB)]
4812                    >\Konr[Descript(T,R5)]
4813                    >\Konf[ordBelt_AdB]
4814                   >Sits[5A_BPrior]=Sits[5A_AdB]
4815                    >E[Ц,e[Ц](5A_BPrior)]=E[Ц,e[Ц](5A_AdB)]
4816                  >elif (XBmax=ReB) then
4817                   >Sits[5A_ReB]
4818                    >E[Ц,e[Ц](5A_ReB)]
4819                    >Konr[Descript(T,R5)]
4820                    >\Konf[ordBelt_ReB]
4821                   >Sits[5A_BPrior]=Sits[5A_ReB]
4822                    >E[Ц,e[Ц](5A_BPrior)]=E[Ц,e[Ц](5A_ReB)]
4823                  >endif
4824                 >Konf[ordBelt2]end
4825                 >Sits[ordBelt]end
4826         >Kont[ordBelt]end
4827         >Kont[ordCore]
4828                 >Sits[ordCore]
4829                  >structure
4830                     >Sits[5A_NiC]
4831                      >E[Ц,e[Ц](5A_NiC)]
4832                       >Sith[T], Fin[T]
4833                        >E[Ч,e[Ч](5A_NiC_T)]
4834                      >Konr[Descript(T,R5)]
4835                       >Sith[T], Fin[T]
4836                       >E[Ц,e[Ц](R5)]
4837                      >Konr[Descript(T,R5)]end
4838                      >Konf[ordCore_NiC]
4839                       >of Konr[Descript(T,R5)]
4840                       >NiC=abs(viality(Konr[Descript(T,R5)]))
4841                      >Konf[ordCore_NiC]end
4842                     >Sits[5A_NiC]end
4843                     >Sits[5A_AdC]
4844                      >E[Ц,e[Ц](5A_AdC)]
4845                       >Sith[T], Fin[T]
4846                        >E[Ч,e[Ч](5A_AdC_T)]
4847                      >Konr[Descript(T,R5)]
4848                       >Sith[T], Fin[T]
4849                       >E[Ц,e[Ц](R5)]
4850                      >Konr[Descript(T,R5)]end
4851                      >Konf[ordCore_AdC]
4852                       >of Konr[Descript(T,R5)]
```

```
4853                              >AdC=abs(viality(Konr[Descript(T,R5)]))
4854                             >Konf[ordCore_AdC]end
4855                           >Sits[5A_AdC]end
4856                           >Sits[5A_ReC]
4857                            >E[Ц,e[Ц](5A_ReC)]
4858                             >Sith[T], Fin[T]
4859                              >E[Ч,e[Ч](5A_ReC_T)
4860                            >Konr[Descript(T,R5)]
4861                             >Sith[T], Fin[T]
4862                              >E[Ц,e[Ц](R5)]
4863                            >Konr[Descript(T,R5)]end
4864                            >Konf[ordCore_ReC]
4865                             >of Konr[Descript(T,R5)]
4866                             >ReC=abs(viality(Konr[Descript(T,R5)]))
4867                            >Konf[ordCore_ReC]end
4868                           >Sits[5A_ReC]end
4869                      >structure end
4870                      >Konr[ordCore1]
4871                       >XCmin
4872                       >XCmid
4873                       >XCmax
4874                      >Konr[ordCore1]end
4875                      >Konf[ordCore2]
4876                       >of XCmin
4877                       >if (XCmax=NiC) then
4878                        >Sits[5A_NiC]
4879                         >E[Ц,e[Ц](5A_NiC)]
4880                         >\Konr[Descript(T,R5)]
4881                         >\Konf[ordCore_NiC]
4882                        >Sits[5A_CPrior]=Sits[5A_NiC]
4883                         >E[Ц,e[Ц](5A_CPrior)]=E[Ц,e[Ц](5A_NiC)]
4884                       >elif (XCmax=AdC) then
4885                        >Sits[5A_AdC]
4886                         >E[Ц,e[Ц](5A_AdC)]
4887                         >\Konr[Descript(T,R5)]
4888                         >\Konf[ordCore_AdC]
4889                        >Sits[5A_CPrior]=Sits[5A_AdC]
4890                         >E[Ц,e[Ц](5A_CPrior)]=E[Ц,e[Ц](5A_AdC)]
4891                       >elif (XCmax=ReC) then
4892                        >Sits[5A_ReC]
4893                         >E[Ц,e[Ц](5A_ReC)]
4894                         >\Konr[Descript(T,R5)]
4895                         >\Konf[ordCore_ReC]
4896                        >Sits[5A_CPrior]=Sits[5A_ReC]
4897                         >E[Ц,e[Ц](5A_CPrior)]=E[Ц,e[Ц](5A_ReC)]
4898                       >endif
4899                      >Konf[ordCore2]end
4900                     >Sits[ordCore]end
4901             >Kont[ordCore]end
4902      >Konq[Rebuild]end}
4903   >Konq[PARTS]end
```

www.ingramcontent.com/pod-product-compliance
Lightning Source LLC
Chambersburg PA
CBHW080828220526
45467CB00008B/2239